A—D

VOLUME 96 ISSUE 1

ARCHITECTURAL DESIGN

EDITORIAL OFFICES
c/o NEW INC
235 Bowery
New York, NY 10002

PRINTER
CONTI TIPOCOLOR S.p.A.

GUEST TYPEFACE
AllCaps | Visual

Printed and Bound in Italy

Available through ARTBOOK | D.A.P.
75 Broad Street, Suite 630
New York, NY 10004
www.artbook.com

COVER IMAGE
Jannes Moons
SPIKA (Insect Perspective Rendering)
KU Leuven, Ghent, Belgium
2025

BIODESIGN IN ARCHITECTURE
GUEST-EDITORS
Rachel Armstrong and Richard Beckett
ISBN: 979-8-9942982-0-6
ISSN 0003 8504

Rachel Armstrong is a pioneering figure in regenerative architecture. As ZAP Professor of Design-Driven Construction at KU Leuven, Belgium, she leads the Regenerative Architecture, Arts & Design (RAAD) research team that investigates the integration of biology, materials science, and architecture. Her work challenges conventional notions of construction by proposing holobiontic architecture—structures conceived as multispecies habitats that participate in ecological cycles. Through innovations in biohybrid materials, such as mycelium composites, unfired clays, and bioreceptive ceramics, Armstrong explores how buildings can grow, adapt, and even self-repair. Her leadership of the EIC Pathfinder project Microbial Hydroponics: Circular Sustainable Electrobiosynthesis (Mi-Hy) exemplifies this vision. Her books include *Soft Living Architecture: An Alternative View of Bio-informed Practice* (Bloomsbury, 2018) and *Experimental Architecture: Designing the Unknown* (Routledge, 2019). Recognized for her visionary and collaborative contributions to sustainable design and innovation, she is a Senior TED Fellow, an EIC Ambassador, and she received a Special Award from the Serbian Academy of Sciences and Arts in 2021. Armstrong's work invites critical reflection on how architecture might evolve in dialogue with the living world. Her research offers not just technical solutions, but a philosophical transition—toward buildings that are not inert shelters but active participants in planetary life.

Richard Beckett is an associate professor in architecture at the Bartlett School of Architecture, University College London (UCL), where he directs RC7 within the BPro (Architectural Design) master's program and co-directs Studio3 on landscape architecture. His research operates at the intersection of computation, biofabrication, and microbial ecologies in the built environment. Internationally known for pioneering probiotic design, a radical architectural approach that advocates reintroducing "good" bacteria into buildings to foster healthier indoor microbiomes, his research was awarded the Design and Technical category and overall President's Medal for Research at the 2020 RIBA President's Awards for Research. Beckett heads the Eng-EPSRC EFRI ELiS: Developing Probiotic Interventions to Reduce the Emergence and Persistence of Pathogens in Built Environments, a multidisciplinary three-year project in partnership with the UCL Eastman Dental Institute and University California San Diego and Johns Hopkins University. His work has been exhibited at *Naturalising Architecture*, the Pompidou Centre (2013), and *Nature—Cooper Hewitt Design Triennial* (2019–20). In June 2025, he co-led the Living Assembly: Building with Biology installation at the London Design Biennale, exploring biologically engineered materials that weave computation and living systems into architecture. Beckett's practice-led research, spanning academia and exhibition, positions him as a seminal voice in reimagining architecture as a living, microbial, and interspecies endeavor.

BIODESIGN
The Living Future of Architecture and Design

Rachel Armstrong and Richard Beckett

Imagine a world where buildings transform our waste into new resources, possess living surfaces that clean the invisible contaminants in the air in our rooms, sustainably grow building blocks rather than extracting them from the ground, and where architectural materials have an innate intelligence that can be digitally accessed so that we can engage with them. This is the promise of biodesign—a movement that blurs the lines between biology and design, between the artificial and the organic.

BIODESIGN IN ARCHITECTURE

Architects have historically been fascinated by the relationships between architecture and nature and, in the mid-twentieth century, biological speculations of "growing" furniture and cities from seeds and spores began to emerge. However, it wasn't until the 1990s that the term "biodesign" first appeared, stemming initially from medical research, where scientists were developing biocompatible implants that could integrate with human tissue. Meanwhile, biotechnological developments were accelerating. Scientists spliced genes, cloned animals, and grew organs in labs. Collaborations with bioartists began to explore these technologies, asking how to design "with" life, while practitioners like Steve Kurtz, founding member of the collective Critical Art Ensemble, framed biology as a medium for cultural critique while questioning the ethics of manipulating life. This fusion of biology and synthetic design laid the groundwork for an increasing DIY approach amongst designers and architects keen to experiment with biological systems.

Design thinking moved to the microscale, scouring laboratory ware for new biological perspectives where the focus was less on morphology, but more about the processes underlying *how* microbes grew. By the early 2000s, designers were actively experimenting with living organisms and materials. In 2012, the seminal book *Bio Design* by William Myers showcased a collective of designers growing bacterial cellulose, considering the processes of decay, cultivating mycelium into molds, and applying slime mold logic—using the organism's ability to find the shortest route between two points, a phenomenon vividly documented in the 2014 film *The Creeping Garden*[1]—to solve urban spatial challenges as living algorithms.

Biodesign has rapidly emerged as a leading research field across design disciplines—from fashion to architecture—through the integration of biological principles and biotechnological innovation. Key exhibitions, such as *Nature—Cooper Hewitt Design Triennial* (2019–20) and *La Fabrique du Vivant* (2019) at the Centre Pompidou, highlight the growing international importance of the field.[2] Its influence continues to expand, driven by a diverse global community and supported by an increasing number of postgraduate and educational programs, like Biodesign Challenge, a design-education platform for high school and college students. This issue of *Architectural Design* offers a critical reflection on biodesign's evolution and its implications for future design practice.

Some of biodesign's most ambitious experiments have remained largely confined to petri dishes and gallery installations, while the first inhabited biodesign house has yet to be realized owing to the many challenges that working with organisms pose. Specifically, architecture continues to grapple with the challenge of scale and commercial integration: Unlike specialized applications in fashion or interiors, architects must contend with structural integrity, building codes, and long-term performance for larger, more permanent applications. A new wave of biodesign pioneers are demonstrating viable commercial pathways for the built environment. These range from biotechnology firms like BioMason (manufacturing biocement tiles), Ecovative (producing mycelium insulation and protective packaging), and MycoWorks (engineering mycelium-based leather and materials) to material science companies such as MOGU (growing mycelium biocomposite for acoustic panels), and research and development firm BIOHM (producing mycelium-based insulation and materials). Their collective work promotes a new design ethos that embraces uncertainty, promotes life, and reimagines buildings as living systems, which in turn necessitates dedicated care, ongoing maintenance, and an acceptance of their unique biorhythms.

This issue of AD argues that architectural innovation will not emerge from laboratories alone, but from design's unique capacity to bridge biological discovery, spatial practice, and cultural engagement. While science excels at quantifying material characteristics—such as mycelium's compressive strength or algae's photosynthetic efficiency—design interrogates how these behaviors can reshape space, experience, and construction logic.

Brought together herein are architects and designers at the forefront of biodesign, identifying four key transitions in contemporary architecture: the metabolic, biodigital, biocomposite, and relational turns. Through diverse and experimental approaches, practitioners are developing a new grammar for sustainable architecture—structures that metabolize, sense, adapt, and evolve alongside their inhabitants, operating as truly living entities.

↰ **Mitchell Joachim, Terreform ONE, and Javier Arbona, Fab Tree Hab, New Windsor, New York, 2025 (model for Natural History Museum of Bern, Switzerland)**

This 3D-printed model captures the latest iteration of the pioneering Fab Tree Hab, a living structure grown from grafted trees and bio-based materials. Conceived at MIT in the early 2000s and now realized as a terrestrial reef, the project redefines architecture as a multispecies habitat, merging computational design with Indigenous agroforestry to combat biodiversity loss.

→ **Kevin Slavin, Elizabeth Hénaff, and The Living / David Benjamin, Subculture: Microbial Metrics and the Multi-Species City, Storefront for Art and Architecture, New York City, 2018**

This installation transformed Storefront's façade into a bio-receptive interface, using sandblasted wood tiles to cultivate and study microbial ecologies unique to Manhattan's Lower East Side. A live metagenomic lab inside the gallery sequenced urban microbiomes, challenging sterile design paradigms and reimagining cities as symbiotic multispecies habitats.

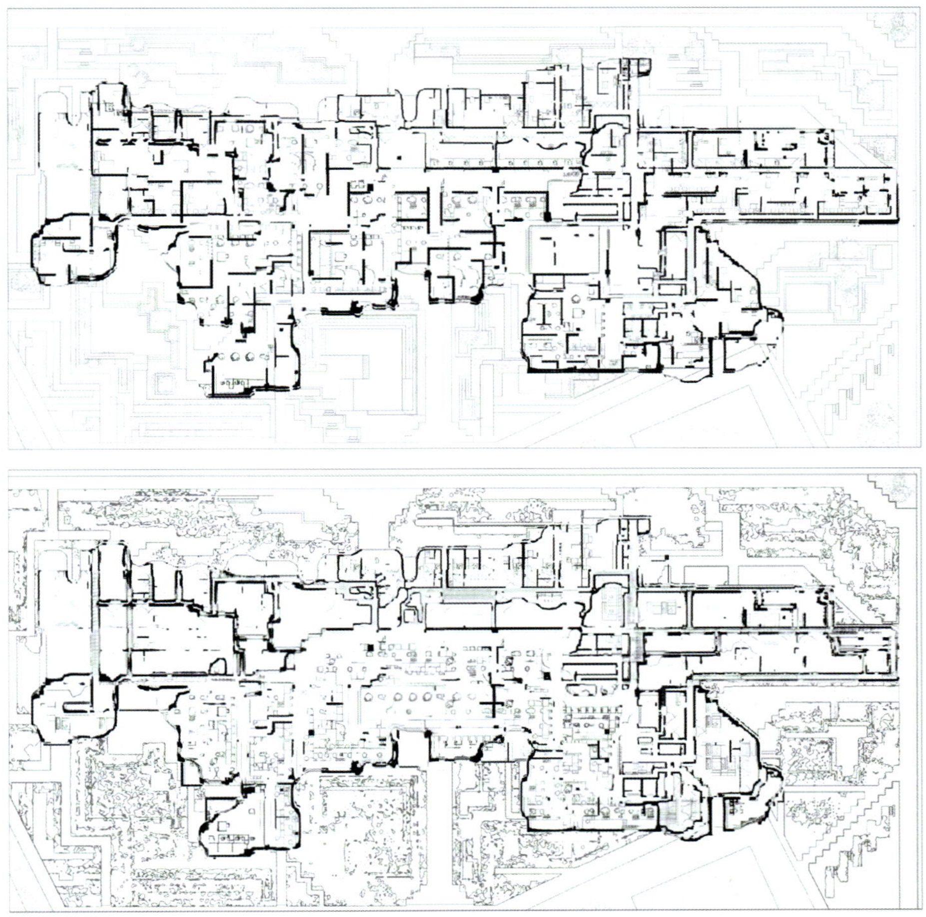

→ **Hangchuan Wei, BioHealer, RC7 / Beckett Lab, London, 2022**

In this biomaterial-driven ageing floorplan, based on generative adversarial networks (GANs), environmental and material ageing data-sets are used to train a GAN model. As materials age and decay, the model reconfigures the program layout. Some spaces remain or are re-grown, while others are allowed to evolve, as indoor inhabitable spaces become outdoor productive spaces. ABOVE: Spatial program plan at six months. BOTTOM: Spatial program plan after eighteen months.

→ **Josefine Ø. Kallehave, Genitrix Nihili: Living Bricks, Paljassaare Peninsula, Tallinn, Estonia, 2017**

These bioactive units, positioned at the threshold of a wastewater plant and nature reserve, metabolize chemical pollutants from seawater into tangible geological records. By materializing environmental data as stratified traces, the "bricks" (or "living" building units) blur the boundaries between remediation, monitoring, and storytelling.

→ **Simone Ferracina, Plug-in Bioreactor Brick(s), Tallinn Architecture Biennale, Estonia, 2017**

Developed with the EU-funded Living Architecture consortium, this modular "brick" integrates microbial fuel cells into buildings, converting wastewater and air into electricity through grape-like clay membranes. The prototype reimagines construction materials as active environmental interfaces, merging bioremediation with energy production.

→ **ecoLogicStudio (Claudia Pasquero & Marco Poletto), GAN-Physarum, Centre Pompidou, Paris, 2022**

This biodigital installation combines living slime mold growth patterns with AI algorithms to reimagine Paris's urban networks. The large-scale "bio-painting" and computational video visualize how biological intelligence can inform resilient blue-green city planning strategies.

THE METABOLIC TURN
FROM BUILDINGS TO BEINGS

A new architectural lexicon is emerging—one that has been expanded beyond static structures to include metabolically active systems. Architecture is undergoing a "metabolic turn," where design becomes a catalyst for ecological processes, embedding biological intelligence to promote holobiontic relationships, interspecies cohabitation, regenerative material flows, and decentralized biomanufacturing. Pioneering this metabolic turn, Rachel Armstrong envisions buildings as multi-organ ecosystems—dynamic entities designed from the inside out. Her "guts-first" methodology begins with internal biochemistry, transforming architecture into a scaffold for life. This approach is encapsulated in projects like SPIKA (2025), a structural pavilion whose bacteriophage-like form—featuring a geometric crown and long spikes—acts as a host for internal microbial metabolisms, bioreceptive surfaces, and hydroponic systems, which are not add-ons but foundational elements. The buildings envisaged by Armstrong support ecological succession, turning them into holobionts—composite organisms where architecture and biology coevolve.

Extending this metabolic logic, Joyce Hwang crafts architectures that forge relationships between species. Her Shelter of Belonging (2024) is not merely a refuge but a multispecies interface, a structure that processes environmental cues and biological needs. With nesting cavities, UV-reflective patterns, and species-specific textures, her work invites nonhuman life into architectural tectonics, where life and death are visible, shared processes within a living system. Phil Ayres explores metabolism through material agency. His biohybrid systems, combining mycelial growth with Kagome weaving, propose architectures in a state of becoming—as structures that metabolize their own materiality. These living composites decay, regenerate, and adapt, embodying temporal agencies that blur the line between building and landscape, structure and organism.

This surrender to metabolic processes finds industrial expression in Ehab Sayed's biomanufacturing company, BIOHM, where materials like Orb and Terraphyll™ are grown from waste streams. These materials do not just sequester carbon, they also synthesize pollutants, support microbial diversity, and demand new metrics for architectural performance. Here, buildings are assessed not by static efficiency but by their ecological processes—their capacity to sustain and regenerate life. Mae-ling Lokko grounds the metabolic turn in cultural practice. Her projects embed agrowaste upcycling into domestic rituals, transforming kitchens into material farms and demolition into nutrient harvesting. Architecture becomes metabolically entangled with everyday life, aligning with natural cycles rather than resisting them.

Together, these practitioners are not merely proposing sustainable alternatives, they are redefining architecture as a metabolic practice. Buildings are no longer constructed but cultivated. They grow, digest, adapt, and decay—participating in the cycles of life. In this future, architecture is evaluated by its metabolic vitality—its ability to sustain, transform, and regenerate the ecosystems it inhabits.

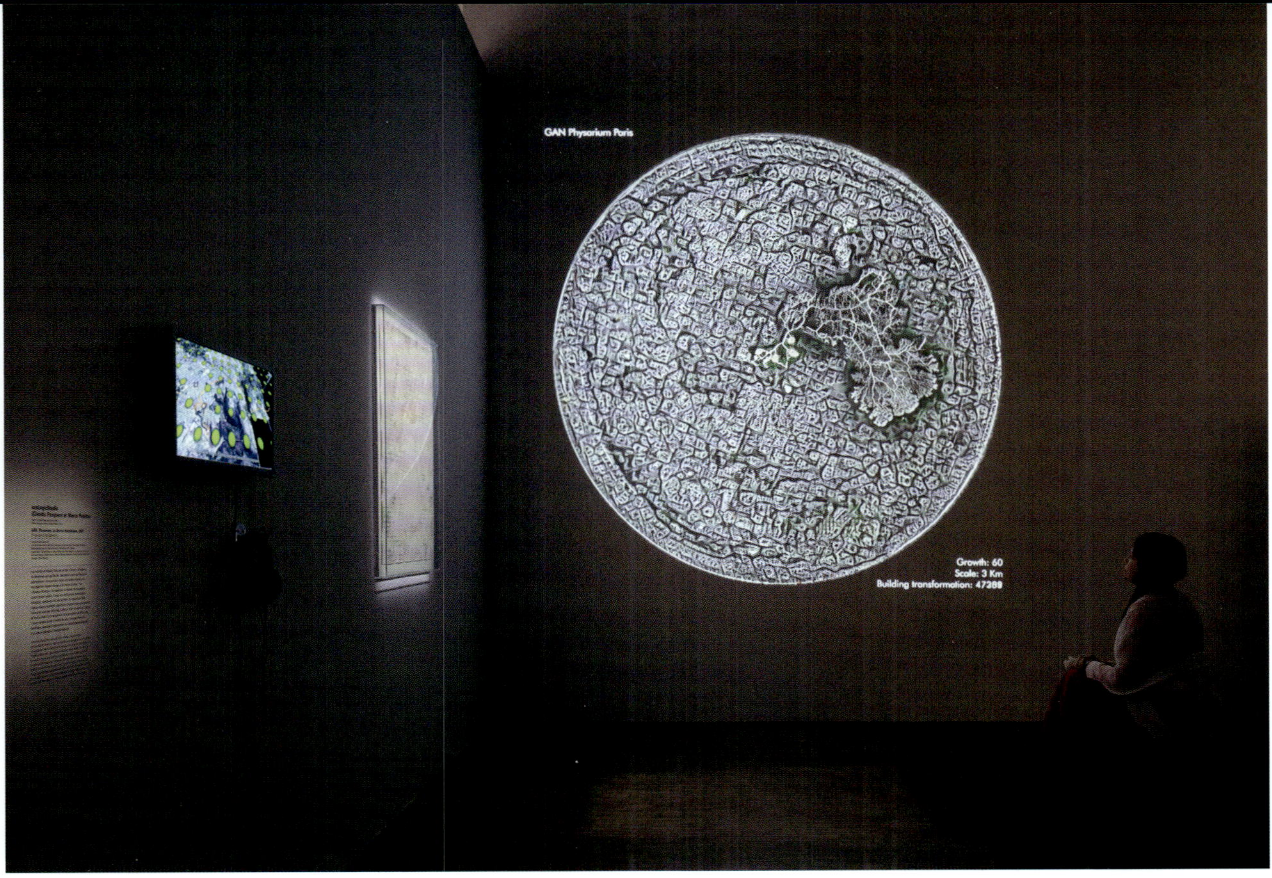

THE BIODIGITAL TURN
INTEGRATING BIOTECHNOLOGY AND INFORMATICS

The convergence of biotechnology and computational informatics is shaping biodesign through an emerging "biodigital turn"—where buildings evolve through a continuous dialogue between digital and biological intelligence. Unlike parametric design, which adjusts fixed variables based on environmental inputs, AI-driven biodigital systems learn from complex biological behaviors. This learned intelligence is then embedded in the building's materials and systems, enabling real-time, unscripted responses to dynamic conditions. This allows architecture to engage with living organisms at scales and speeds beyond human perception.

Deep.Forest (2024) by ecoLogicStudio (Claudia Pasquero and Marco Poletto) exemplifies this approach, treating architecture as a cybernetic organism. Satellite imagery is processed through mycelium-inspired algorithms to generate spatial configurations, while living walls of algae and fungi metabolize carbon dioxide and urban waste. AI systems monitor and regulate these biological processes, optimizing conditions for both microbial life and human comfort. In projects like GAN-Physarum (2022), machine learning generates urban plans based on biological growth patterns, demonstrating how computation can interpret and extend the logic of living systems.

At [EmTech], biological behaviors are abstracted into algorithmic design frameworks. The graduate program's "genome-to-G-code" methodology translates natural phenomena—such as mycelium's branching logic or bamboo's hygroscopic movement, where its dimensional changes are directly driven by the water-absorbing properties of the material—into computational rules. These are embedded into robotic fabrication systems that respond to real-time sensor data, enabling materials to adapt during construction. Projects like Gap in the Cloud (2024) scale these principles to the city, treating urban environments as metabolic networks of energy and nutrient exchange. Kyoung Hee Kim advances this integration through AI-optimized microalgae façades that function as both climate regulators and biofuel generators. Her Biochromic Window (2021) dynamically adjusts to environmental conditions, while producing oxygen and biodiesel. In this instance, AI continuously fine-tunes growth parameters, creating closed-loop systems that operate across multiple timescales.

Together, these practices establish a biodigital platform where AI acts as an extended autonomic nervous system—quietly interpreting sensor data, regulating biological processes, and enabling design through subliminal forms of communication. This is not automation replacing nature, but computation deepening our collaboration with it.

THE BIOCOMPOSITE TURN
EMERGING BIOMATERIALS

The emergence of living biocomposites represents the "biocomposite turn"—a significant development in architectural thinking, where materials are no longer merely shaped but grown. This transition repositions the designer as a cultivator of biological processes. A new generation of architects is developing materials that are not only low carbon but regenerative, productive, adaptive, and alive. Unlike conventional composites, these living materials are coproduced with microbial partners, whose growth patterns, pigmentation, and structural behaviors shape the final form. The result is a new class of building skins, interior finishes, and even structural systems that evolve over time—molding themselves into spaces, responding to environmental stimuli, and participating in ecological cycles. Yet the full metabolic range of this field is only beginning to be explored: The media that support growth, the scaffolds that guide it, and the systems that sustain it are all active sites of experimentation and future research.

The work of Marcos Cruz and Brenda Parker exemplifies this ethos. Their bioreceptive concrete and ceramics are engineered to host mosses, lichens, and algae, transforming façades into living filtration systems. In Augmented Polycultures (2024), environmental data guides microbial colonization, turning walls into habitats that purify air and water while supporting biodiversity. bioMATTERS (Nancy Diniz and Frank Melendez) regard microorganisms as creative collaborators in their designs. In their MYCO-ALGA tiles (2022–23), mycelium and algae intertwine with 3D-printed scaffolds, and biological growth becomes a design gesture. Rather than imposing form, machine learning adapts to these organisms' behaviors, resulting in a negotiated architecture—one that emerges, adapts, and evolves in dialogue with its microbial co-authors. Maria Kuptsova's ARBOR project (2020–) develops these concepts further, embedding biosensors into mycelium-infused wood waste to create structures that mimic forest ecologies. Her installations respond to environmental changes, redistribute nutrients, and repurpose decay as growth.

Architects exploring the biocomposite turn are not just inventing new materials, they are redefining how our buildings are constructed. In this paradigm, architecture becomes a living interface for microbial builders, where success is measured by the capacity of the architect to develop frameworks that sustain, advance, and regenerate life.

THE RELATIONAL TURN
CHALLENGING PERMANENCE AND CONTROL

The "relational turn" in architecture marks a change from buildings as fixed, inert objects to understanding them as responsive, evolving participants in ecological systems. This perspective dissolves the boundary between built form and living environment, inviting a deeper entanglement between human and nonhuman life. Neil Spiller describes a series of projects undertaken at Hooke Park in Dorset, UK, that utilize the nonuniformity of natural forms, expressing qualities of "wolfish life." Biodesign welcomes impermanence, uncertainty, and material vitality. It resists the modernist impulse to dominate nature, instead cultivating porous, adaptive architectures that grow, decay, and regenerate. In this view, buildings are no longer pristine objects of control but messy, symbiotic agents cohabiting with microbes, weather, and time, attuned to rhythms beyond our own. In letting go of control, architecture finally becomes alive—less structures to be maintained than spaces to be tended and nurtured.

Richard Beckett's pioneering Probiotic Tectonics (2022) challenges the sterility of modernist architecture by designing buildings as "living immune systems." His eroding TerraSolum walls (2023) defy the pursuit of permanence, instead cultivating microbial ecosystems that reflect our own microbiomes. In this vision, buildings become extensions of the holobiont body, their porous surfaces exchanging microbes with inhabitants, their material decay contributing ecologically. Architecture here is not static but symbiotic, with human health deeply connected to a building's capacity to host and sustain microbial diversity. CITA (Mette Ramsgaard Thomsen and Paul Nicholas) extend this logic to material life cycles. Their Repair Ecologies project (2025) redefines durability as the capacity to evolve with it. By using biopolymers that can be robotically reshaped with heat and humidity, they create architectures that *learn* from local ecological cycles. A wall might be repaired with eelgrass harvested in spring or mussel shells collected in winter—its lifespan measured in seasons rather than centuries. This is architecture as a *collaborative multi-agent practice*, where maintenance becomes a dialogue with place, inhabitants, and time.

← **Miruna Porosnicu, Roba Abdelhak, Yuchun Lu, Wei Zhang, Iravati Wagle, and Qingxuan Li, B.L.U.R, RC7 / Beckett Lab, London, 2025**

B.L.U.R explores the application of grown "engineered living materials" for architecture. Given its moisture-dependent nature, bacterial cellulose-based materials challenge conventional notions of inertness in architecture, instead exhibiting animate material properties of self-healing and environmental responsiveness. Digital technologies are integrated into biomaterial sensing workflows to monitor the cycling of material and living metabolic properties.

← **Can Yardimci, Zhiyuan Wu, Ziyi Liu, Qing Wang, and Yaoyao Yang, BioTessera, RC7 / Beckett Lab, London, 2025**

BioTessera explores a new paradigm of architecture where ageing, repair, and ephemerality are no longer seen as failures but as opportunities for adaptation and resilience. Mycelium ageing experiments train an AI image detection model for rapid monitoring of mycelium ageing, in real time on buildings.

1 Nicolas Rapold, "'The Creeping Garden,' on the Wonders of Slime Mold," review of *The Creeping Garden*, directed by Tim Grabham and Jasper Sharp (Arrow Academy, 2014), *New York Times*, September 29, 2015, https://www.nytimes.com/2015/09/30/movies/review-the-creeping-garden-on-the-wonders-of-the-slime-mold.html.

2 *Nature—Cooper Hewitt Design Triennial*, May 10, 2019–January 20, 2020, Cooper Hewitt, New York; and *La Fabrique du Vivant*, February 20–April 15, 2019, Centre Pompidou, Paris.

Jonathan Dessi-Olive explores architecture's mortality through mycelium bricks, confronting an uncomfortable truth: Genuine sustainability embraces decay. By designing for decomposition—using hang-drying techniques that let fungal networks shape the form—he creates structures that are designed to disappear. Their collapse is not failure, but is intentional—a return to the earth. In this approach, deterioration, decay, and decomposition become *ecological generosity*, as buildings gift nutrients back to the soil. Ronald Rael reimagines the "relational turn" through the lens of deep material time, treating clay as an active ecological agent. His 3D-printed adobe structures—like Casa Covida (2019) and Skylos (2022)—merge Indigenous earth-building with robotic precision, embodying a paradox of the ancient and algorithmic. Rael's architectures reveal how materials that are conventionally considered inert can metabolize, possess an innate intelligence, and host life, resisting extractive construction logics. His buildings are not just biodegradable but biographical, recording weather, labor, and cultural repair. In doing so, Rael expands relational design to include geological and cultural timescales, positioning architecture as a custodian of memory and producer of future soil.

Together, these designers propose a significant cultural transition: from sterile spaces that we ignore as backgrounds to our lives, to circular, living systems we care for. Their work imagines a future where wiping a fingerprint from a wall seems as absurd as sterilizing a garden—where architecture is not separate from us, but part of us. In turn, we become part of our living spaces. Imperfect, evolving, and vigorously alive, our buildings are not merely habitats but bodily extensions.

A NEW SYNTHESIS

The metabolic, biodigital, biocomposite, and relational turns in biodesign are not just redefining architecture, they are reanimating it. No longer inert or backgrounded, buildings are emerging as living organ systems: metabolically active, materially intelligent, and ecologically entangled. The designers in this issue of *Architectural Design* reveal early glimpses of this transformation. Their work directly answers the call by thinkers like Beatriz Colomina and Mark Wigley to decenter the human and embrace a "biotic architecture," which acknowledges that we have always been, and will always be, entangled with microbial life. As they argue in these pages, the modernist "antibiotic attitude" that sought to sterilize the built environment has produced a silent extinction of microbial diversity, directly implicating architecture in contemporary health crises. The projects within this issue propose an alternative: architecture as a responsive, coevolving entity—one that breathes, metabolizes, digests, senses, and thinks. This is not architecture as a prophylactic barrier but as a porous, curatorial interface, a built extension of the gut microbiome that actively promotes ecological exchange and health. These structures are not merely shaped by humans; they shape us in return. Worn like sticky biofilms, they extend into ecosystems, rethreading us into the cycles of life. This is architecture as more-than-human: a synthetic skin, a distributed intelligence, a negotiated cohabitation between body and world. It challenges us to regard it as an active participant in the messy, endlessly creative web of life.

Rachel Armstrong

Biodesigning Cities from the Guts First

→ **Rachel Armstrong and team, Living Architecture, University of the West of England (UWE), Bristol, UK, 2019**

A detailed section of the Living Architecture wall, featuring three interconnected bioreactors: microbial fuel cells, photobioreactors, and synthetic bioreactors engineered to extract phosphates from detergents. Sustained by domestic waste water (kitchen, bathroom, and urine) and sunlight, and producing bioelectricity, cleaned water, and biomass, the system operated as a closed loop, demonstrating how architecture can integrate metabolic processes.

→ **Rachel Armstrong and team, Active Living Architecture: Controlled Environment, Chaleur Humaine Art and Industry Triennial, Dunkirk, France, 2023**

An autonomous microbial "cyborg" that sustained itself through waste processed by its microbial fuel cells. The installation communicated with viewers by converting microbial fuel cell action potentials into digital animations, reflecting microbial "comfort" while allowing human interaction via remote feeding and temperature adjustments.

Biodesign must evolve beyond the design of materials and surfaces—or risk becoming complicit in the Sixth Great Extinction. Of the 39 percent of global carbon emissions generated by the practice of the built environment, 28 percent comes from building operations while only 11 percent is from materials and construction.[1] Beyond providing compelling aesthetics and new materials, biodesign needs to perform the work of ecosystems in our living spaces—processing local waste, purifying water, filtering air, and exchanging energy in smart, circular loops. Through rethinking our whole approach to building systems, which are currently centered on fossil fuel and waste generation, architecture can transform spent substrates into resources, creating closed loops where materials are recycled and even regenerated into valuable new configurations. Through adopting a visceral "guts-first" approach to building service design, our living spaces can support our daily lives, transforming into self-renewing, waste-upcycling, and energy-autonomous systems—just like nature itself.

Forget burning fossil fuels. The ultimate clean-tech revolutionaries are microbes. They restore environments, building biodiversity as they work. Bacteria, fungi, microalgae, and their microscopic kin have spent billions of years perfecting what human engineers struggle to replicate: zero-waste energy, self-repairing systems, and closed-loop chemistry. As Nature's original architects, microbes use nanotechnology in their powerful metabolisms to achieve their design goals—and they're ready to dismantle the petroleum age.

Biodesign is working with microbes to impact the future of building performance, to turn waste into wealth: Sewage becomes fertilizer, greywater generates electricity, and walls turn into carbon-hungry, air-scrubbing skins. Buildings are no longer just structures, they are living metabolic hubs—whose interior microbiomes echo the functions of a gut—where every input is recycled and every output is fuel or the starting point for another living process.

Prototyping
Metabolic
Architectures

↙ **Cécile B. Evans and Rachel Armstrong, 999 years 13 sqm (The Future Belongs to Ghosts), Whitechapel Gallery, London, 2019**

999 years 13 sqm presented an architectural ecosystem devoid of human occupants, sustained solely by microbial life and maintained through weekly human intervention in the gallery. The work explored posthuman architecture, where digital ghosts and microbial agency replace traditional habitation, framing autonomy as a collaboration between nonhuman actors and lingering protocols of human care.

→ **Research Team, Microbial Hydroponics (Year 1 prototype), University of Southampton, UK, 2025**

In this project, hydroponic basil plants were nourished by microbial fuel cell biofilms, which replaced synthetic nitrates derived from the greenhouse gas-intensive Haber-Bosch process. The biofilm in the microbial fuel cells mimicked a plant's rhizosphere, regulating growth and producing flavor through biochemical interactions fueled by household wastewater—offering a sustainable alternative to conventional hydroponics.

The potency of microbial metabolism is the proven bedrock of a post–fossil fuel future, not science fiction. Pioneering work in metabolic biodesign cultivates microbial communities to turn household waste into renewable resources, on-site and in real time. Like a smart, living tapestry, these systems transform waste—their "food"—into useful household resources, from cleaned water to bioelectricity and biomass. These prototypes are already here, with designs that begin by weaving microbial networks into the very anatomy of homes—embodying a *guts-first* approach to architecture.[2]

This vision of buildings as "living" metabolic entities was demonstrated in the Living Architecture project (2016–19), funded by the European Union's Horizon 2020 Research and Innovation Programme, to reimagine buildings as symbiotic superorganisms.[3] The project cultivated discrete populations of microbes—photosynthetic, anaerobic, and synthetic communities—in different but interconnected bioreactors where they performed specialized metabolic functions, much like the chambers of a cow's stomach. By plumbing these bioreactors together and coordinating them via bioelectricity, the project closed resource loops entirely to create a dynamic, self-regulating system. The waste from one process became fuel for another, generating self-powering bioelectricity in the process. As different microbes can deploy various metabolisms, the specific outputs for each bioreactor component could be selected and altered. The system, therefore, comprised a modular, programmable living metabolism that was constantly self-powering, digesting, transforming, and producing biomass.

For such systems to be accepted and to thrive, people must want to engage with them—a challenge when microbes are invisible and "slimy" to the untrained eye. The Active Living Infrastructure: Controlled Environment (ALICE) prototype (2019–21), funded by the EU's Innovation Award scheme, made a critical breakthrough in metabolic biodesign. The project demonstrated that microbial "comfort"—encoded in electroactivity—can be monitored and interpreted, enabling residents to observe and optimize their living building systems in real time by adjusting bioreactor conditions. The UK-based ALICE consortium (Newcastle, UWE Bristol, and Translating Nature) allowed real-time bioreactor monitoring, transforming real-time microbial signals into animations that enabled human-microbe "dialogue." This development was foundational for scaling metabolic architectures and building broader acceptance for microbial systems capable of turning buildings from inert structures into responsive, semi-biological entities.

Could architectural microbes outgrow human oversight? The collaborative project 999 years 13 sqm (The Future Belongs to Ghosts), with artist Cécile B. Evans, envisioned electroactive biofilms as permanent architectural residents in self-sustaining bioreactors. Humans faded to digital ghosts in this microbial future, developed by UWE Bristol, Edinburgh, and Newcastle universities for Whitechapel Gallery's 2019 *Is This Tomorrow?* exhibition. Here, buildings achieved true autonomy—thriving without human intervention.

Autonomy, however, does not preclude complexity. The Microbial Hydroponics: Circular Sustainable Electrobiosynthesis (Mi-Hy) project (2023–27) advances microbial metabolism for buildings even further, by replacing chemical fertilizers in hydroponics with natural biomolecules generated and regulated by an electroactive "prosthetic rhizosphere." Led and coordinated by an EIC Pathfinder-funded consortium of universities,[4] together with the Spanish National Research Council, Sony Computer Science Laboratories (France), and Biofaction (Austria), the Mi-Hy system transforms wastewater into food biomass, clean water, bioelectricity, fertilizers, and self-assembling bio-bricks to generate an endlessly productive resource loop.

Powering urban services with living metabolisms, these prototypes pioneer a regenerative infrastructure paradigm. Microbes purify air, exchange electrons, and convert waste into resources, enabling cities to escape petroleum dependence through circular economies where waste nourishes new growth.

The Structural Protection for Interdependent Holobiont Assembly (SPIKA) installation is the latest demonstrator of these principles. This autonomous microbially-enabled urban agent is based on Mi-Hy science and was created for Mark Wigley and Beatriz Colomina's *We the Bacteria: Notes Toward Biotic Architecture* exhibition at the 24th Triennale Milano, Italy, in 2025. Its thorny exoskeleton and bioreceptive panels formed a defensive interface for its microbially powered core—an architectural carapace that protected, provoked, and hosted life. A self-sustaining Mi-Hy ecosystem thrived within: *Geobacter* biofilms generated electricity to power LED grow lights while simultaneously producing biologically available nitrogen to nourish plant roots. Here, electroactive biofilms and hydroponic plants formed a closed nutrient loop—an intimate exchange of energy and sustenance, orchestrated under a controlled UV spectrum—as a crucible for microbial intelligence and vegetal life.

Externally, SPIKA's bioreceptive surfaces initiate a slow-motion vertical ecological succession. Algae colonize the structure's skin as pioneer species, giving way to lichens, mosses, and eventually complex flora, through designed facilitation. The defensive morphology is functional, enabling an uninterrupted ecological evolution—from biofilms to rootless plants and thriving, adaptive ecosystems that attract insects and opportunistic parasites, forming an *ecology of respect*, where human rituals of care replace extraction and control.

SPIKA acts as a living prosthesis embedded in the urban fabric, pioneering a new model of coexistence, which is earned through mutuality and care. Here, wildness is a necessary condition for post-anthropocentric architecture, casting it as a protector, mediator, and provocateur in the ongoing negotiation between human ambition and nonhuman autonomy.

In these prototypes, metabolic design begins from inside the workings of the space and spreads outwards through the development and installation of architectural *viscera*. Microbes act as natural sensors and effectors, detecting and

← **Rachel Armstrong and team, Guts Out (Bioreceptive Panels with Humidifier Organ), 24th Triennale Milano International Exhibition, Milan, Italy, 2025**

Bioreceptive panels with villi-like protrusions were seeded with microorganisms to maximize the surface area for microbial colonization and ecological succession, transitioning from biofilms to rooted plants. The design referenced intestinal villi to optimize nutrient exchange and the formation of microbial habitats, and to promote biodiversity, while integrating a humidifier to provide moisture for sustaining their local microclimates.

← **Rachel Armstrong and team, SPIKA (Built Structure), 24th Triennale Milano International Exhibition, Milan, Italy, 2025**

A next-generation architectural core comprising microbial fuel cell arrays, hydroponic towers, and bioreceptive panels. Its bacteriophage-like morphology metaphorically aligned with the insertion of genetic programs into host architecture—the structure's "spikes" suggesting a transfer of ecological technology into buildings, much as viruses deliver genetic material into host cells.

→ **Jannes Moons, SPIKA (Insect Perspective Rendering), KU Leuven, Ghent, Belgium, 2025**

A rendering of SPIKA as perceived by insect vision such as a bee's, depicting UV-enhanced color and electromagnetic fields. Designed as a pollinator attractor, the structure's form and surface chemistry aimed to increase urban biodiversity by providing sensory experiences for nonhumans.

responding to environmental changes; a potency that can be harnessed through digital communication systems. This *guts-first* approach is founded on introducing life-supporting microbial exchanges and chemical flows into building systems—via electron transfers between bacteria, symbiotic waste transformation, and nutrient cycles—to gain a bidirectional exchange between environmental data and biological processes, analogous to a gut/brain axis. Here, the "gut" represents the metabolic activity of microbes, while the "brain" symbolizes the building's digital network that processes and reacts to this information, not as a literal brain but as a responsive control system. Architecture becomes an adaptive, living system where microbial intelligence and digital infrastructure work in concert to optimize building performance and sustainability. These microbial-mediated systems are rich with life-supporting flows of oxygen, water, and energy. Functioning from a microbial core to a living surface, layer by layer, they enable buildings, microorganisms, and all inhabitants to thrive.

← **Rachel Armstrong and team, SPIKA's Parasite (detail), designed by Jannes Moons, 24th Triennale Milano International Exhibition, Milan, Italy, 2025**

A gut-like, parasitic extension of SPIKA that hijacked host structures to propagate its living building systems. Embodying the notion that parasites are *noise-makers*—agents that disrupt systems to create new exchanges—it acted as an architectural "interrupter," forcing host buildings to adapt by integrating microbial life to establish a generative, symbiotic relationship.

→ **Jannes Moons, SPIKA (Insect Perspective Rendering), KU Leuven, Ghent, Belgium, 2025**

Two core principles drive this approach. First, *systems design* positions the architect as a metabolic engineer, choreographing interactions among microbes, plants, and humans. Wastewater powers microbial fuel cells; plant roots stabilize living walls; human presence sustains the cycles that bind these systems together. Second, a *cyborgian logic* guides architectural evolution. Buildings adapt to environmental changes by forming new metabolic layers—photosynthetic cladding, bioreceptive skins—around foundational systems. This technological adaptation mirrors biological evolution, where ancient structures like the gut are encased by adaptive segmental innovations like lungs or limbs. Crucially, aesthetics emerge from *ecological success*—from systems that survive, adapt, and flourish—rather than from obeying imposed human order. A moss-covered wall, for example, is beautiful because it has negotiated its survival and optimized its conditions for life.

Reimagining cities as *constructed wildernesses*, hybrid ecosystems like SPIKA act as protective prostheses for nonhuman life. Instead of taming nature, these architectures defend its autonomy. Microbes, plants, insects, and humans coexist as integral to living systems that do not rely on constant human control but on *reciprocal care*.

Ultimately, this metabolic biodesign reframes architecture as an act of ecological partnership. By grounding buildings in microbial processes and material flows, we extend metabolic systems into the landscape. Here, human habitats reinforce rather than disrupt nature's cycles, ensuring mutual survival through daily rituals of care, where energy circulates, diversity builds resilience, and every output becomes an input. Realizing this vision requires architects to collaborate with nature's microbial engineers by becoming fluent in their languages—decoding bacterial electricity, fungal networks, and symbiotic exchanges—to develop tools, materials, and spatial protocols for interspecies cohabitation.

Design must perform the *work of life*,[5] by inviting living processes into our spaces, ensuring technologies respond to place-based resource flows, and allowing aesthetics to emerge from ecological function. Think of the grotesque magnificence of the anglerfish's bioluminescent lure, or mycelium's stealthy digestion of organic debris. Designing from the *guts* out, the metabolic architect becomes a negotiator and translator of microbial intelligence—defending space for life to flourish within walls, beneath floors, and across rooftops. These architectures unsettle because they flourish *despite* human aesthetic norms. Here, the architect is guided by life-centric protocols before stylistic intent.

The resultant spaces are not "machines for living in"[6] but rather living treaties: photosynthetic skins harvesting light into biofuel, fungal networks digesting waste into structure, bacterial communities regulating air like microscopic lungs through the two-way exchange of a gut/brain axis. The material changes may seem slight—a bioreceptive façade, a living solar panel, an electroactive film—but their cumulative impact is transformational. Where once we imposed rigid systems on nature, we now design using cooperative strategies, establishing mutualistic routines between humans and the environment.

Toward an Architectural Gut/**Brain Axis**

1 World Green Building Council, *Bringing Embodied Carbon Upfront* (London: World Green Building Council, 2019), https://worldgbc.org/climate-action/embodied-carbon/.
2 Rachel Armstrong, "The Body as an Architectural Space—From Lip to Anus, the Gastrointestinal Tract as a Site for Redesigning and Development," in "Integrating Architecture," ed. Neil Spiller, *Architectural Design* 66, pp. 9–10; 86–91.
3 The project was a collaboration of experts from the universities of Newcastle (UK), the West of England (UWE Bristol, UK), and Trento (Italy), alongside those from the Spanish National Research Council, Liquifer Systems Group (Austria), and Explora Biotech (Italy).
4 The universities include KU Leuven (Belgium), Southampton (UK), and the UWE Bristol (UK).
5 Rachel Armstrong, "Life, Mind and Matter: Chemistry for an Ecological Era," *Social Epistemology* 38, no. 6 (2024): 774–84.
6 Le Corbusier, *Toward an Architecture*, trans. Frederick Etchells (London Architectural Press, 1927).

Images: p. 13 (top) Courtesy Living Architecture consortium. Photo: © Rachel Armstrong; p. 13 (bottom) Courtesy ALICE consortium. Photo: © Rachel Armstrong; p. 14 Courtesy of the artists. Photo: © Cécile B. Evans; p. 15 Courtesy Microbial Hydroponics consortium. Photo: © Microbial Hydroponics consortium; pp. 16, 18 Photo: © Işıl Yücel, Spika consortium; pp. 17, 19 Rendering courtesy of Jannes Moons, Spika consortium. Image: © Jannes Moons

This is the *Visceral Turn* in biodesign, an architecture that begins with the microbial exchanges and fungal circulations that have sustained life for 3.8 billion years. Life-promoting protocols—in the form of metabolic flows, material loops, and adaptive resilience—once shaped every ecosystem. The Anthropocene's error was to overwrite this operating system with extractive infrastructures, concrete and steel demanding unnatural energy inputs. Now, as climate collapse reshapes the rules of habitation, architects must reinstate nature's building code as the basis for all design. Here, architecture is inverted: Biological processes shape form and function to become regenerative organ-complexes embedded in living landscapes. The architect is a metabolic curator, orchestrating flows of carbon, water, and energy to sustain multispecies life—and provoking surprising new aesthetics.

In this paradigm, buildings evolve into autonomous, life-bearing systems: They filter, repair, and regenerate as naturally as they decay. As original expressions of nature, cities are uniquely adapted for human cohabitation—yet operating on life's terms. Every architectural surface and system become self-sustaining, infused with intelligence and ecological purpose. The result is a new kind of urban landscape: an integrated ecology, where human culture, building systems, microbes, "living" materials, wilderness, and planetary cycles individually negotiate their position within a collective network that comprises the municipal system. Like all sufficiently advanced infrastructure, the Visceral Biodesign city achieves the inevitable, a building technology so thoroughly evolved it disappears into nature's operating system.

ON BELONGING

CREATING SPACES FOR MULTISPECIES CARE

Joyce Hwang

Cradled in the rolling hills of Umbria, Italy, the Civitella Ranieri artist residency program offers a rare blend of creative sanctuary and living history. Housed within a fifteenth-century castle—once a fortified stronghold and home to the Ranieri family—the residency invites visual artists, architects, musicians, and writers to immerse themselves in a space where past and present converge. Here, medieval stone walls and sweeping countryside vistas frame a supportive, interdisciplinary environment, designed to nurture innovation while echoing the legacy of a place built for both protection and community.

↑ **Joyce Hwang, Shelter of Belonging, Civitella Ranieri, Umbertide, Italy, 2024**

Shelter of Belonging, a multispecies habitat installation, is sited at the edge of the forest on Civitella Ranieri's grounds. The structure provides a space of respite in the woods for Civitella's visitors and residents, while maintaining a visual connection to the fifteenth-century castle.

Shelter of Belonging is a permanent multispecies habitat created during the summer residency of 2024 to provide havens for local fauna—including bats, birds, and solitary bees, as well as smaller terrestrial animals. Sited at the boundary between Civitella's forest and its more manicured landscape, the design itself alludes to liminal spaces on Civitella's grounds, where wildlife dwells in close proximity to human life.

Unlike the concept of home—which is associated with a feeling of belonging—the word "shelter" has a more complicated meaning. It's often tied to urgent or temporary situations, like emergency shelters for crises, fallout shelters for disasters, refugee shelters for displaced people, or transitional housing for those in need. It can also refer to places that take in strays or abandoned animals, like an "animal shelter." However, "shelter" isn't just about urgency—it can also suggest a safe space to pause, rest, or escape from the stresses, routines, or complexities of daily life. While the habitat installation was conceived of as a shelter providing respite for both humans and nonhumans, it also serves as a shelter that addresses the urgencies of our planet's ecological crises, including biodiversity loss and habitat destruction.

Designing **for Human Users**

Given the opportunity to design and build an installation at Civitella, one of the primary challenges was to create something that responded to the lives and habits of temporary visitors within a setting that has been cultivated over centuries of care. Within the intimate, inwardly focused context of the residency, the nuances of daily life, of both resident humans and nonhumans, became apparent. The manicured lawn was seen as an open space for socialization, while the woods were a place to retreat to, with some residents taking private phone calls standing on the wooded path.

Located near this boundary, just slightly outside of the designated social space of the gazebo and lawn chairs, the Shelter of Belonging's wood-framed structures provide benches for human residents to rest and relax, arranged to enable different social scenarios. The structure and seating are built from locally sourced larch wood, crafted with the support of FAGI Infissi, a nearby wood fabrication company. Habitat boxes are created with local cedar, as well as salvaged wood collected from the grounds by Civitella's groundskeeper, Marco Alunni. An additional seat for humans is constructed using a gabion filled with salvaged building fragments, located underneath the structure—serving additionally as habitat for ground-dwelling animals.

By introducing seating catered to humans in an installation designed for wildlife, the project intends to bring human visitors closer to the animals inhabiting the structure, as well as giving a platform for experiencing the sounds and sights of animals within the woods. This emphasis on creating a sense of physical proximity is also reinforced through the tactility of the wooden habitats, as they encourage human visitors to stop, observe, and perhaps even to touch.

→ **Sleeping bat at Civitella Ranieri, Umbertide, Italy, 2024**

The Civitella Ranieri castle and grounds are already filled with abundant instances of multispecies encounter. This image shows a sleeping bat in a door jamb of the gallery and lecture space, discovered by Civitella Ranieri staff and residents.

→ **Joyce Hwang, Shelter of Belonging, Civitella Ranieri, Umbertide, Italy, 2024**

The wood-framed structure is enhanced with various habitat boxes, designed by digitally scanning castle and garden walls and using the scans as source material for modeling new habitat spaces. The boxes are both CNC-milled and hand carved.

Animal encounters already happen on a daily basis at Civitella. Lizards consistently scurry across walkways and steps, disappearing into crevices in the ground and castle; residents even encounter them inside their apartments. Bats can be seen in multiple places: inside an apartment's washroom, behind a door jamb, behind a picture frame, and between structural beams. Bat guano is found in various studio spaces and in shared rooms as well. Pigeons perch on balconies. Civitella's human residents seem to have mixed reactions to animals in and around their living spaces, with some expressing intrigue, and others voicing discomfort and calling in staff to help remove these animal occupants.

Within this context of semi-regular animal encounters, how might an outdoor installation heighten the curiosity that humans have toward wildlife, or perhaps soften the unease they may feel when animals are too close? Attentive to such questions, Shelter of Belonging establishes circumstances for staging potential, fleeting encounters between humans and animals in a kind of constructed third space between two worlds—neither in a human's controlled living quarters nor in the lesser-maintained woods or vacant structures.

Designing **for Nonhuman Species**

Another challenge at Civitella was to consider how to design for, and with, the animals who have lived and thrived on the grounds for generations. As an outsider, how might one import knowledge and ideas to a place that already has a long history of culture and ecology? How can one "read" an environment to understand it and its constituents? How can an architectural design process be a form of collaboration with the built and living environment?

A previous resident in 2012 had documented that specific birds were "nesting in castle / castle walls," a notation that led to an exploration of the castle's walls and to the animals who live on the margins within close proximity to human spaces. A variety of wildlife recording equipment was deployed, including an ultrasonic bat detector, an acoustic recorder focused on birds and frogs, as well as cell phone apps such as Merlin Bird ID. Between June 13 and July 3, 2024, these collected 194,427 recordings of birds and frogs, and an initial analysis revealed around 244 species identified by different calls. Some 20,550 instances of bats were also recorded. The grounds seemed to be home to more bat species than expected: In comparison with the nine species of bats in New York State, the recorders logged nineteen species of bats in Umbertide. The most frequently noted was the common pipistrelle, with almost 7,000 recordings (6,769 recordings out of the 20,000 or so). The data collected even indicated the possible presence of a few endangered or rarely spotted bat species as well.

→ **Joyce Hwang, Shelter of Belonging, Civitella Ranieri, Umbertide, Italy, 2024**

Habitat boxes, inspired by the varied surfaces and forms of the Civitella Ranieri castle walls, adorn the wooden structure. This image illustrates a close-up view of the machine-milled habitat boxes.

The grounds of Civitella offered a wealth of clues about the specific environments—or *umwelt*, as described by biologist Jakob von Uexküll[1]—of host animals. A lot of time was spent walking around the property, making close observations, and following hints. This included examining the text messages of fellow residents about animal sightings as well as notes written by previous residents, to identify locations within the castle walls, garden structures, and other spaces that evidenced animal occupation. Experimenting with ways of digitally scanning these spaces using Polycam, these scans were then translated into digital models to help develop a spatial vocabulary for wildlife habitats, including crevices, holes, cavities, cracks, and layers, among other conditions. The scans themselves—digital interpretations of the castle walls and surrounding structures—became the source material for modeling a variety of habitat boxes for solitary bees (who live in deep holes of various sizes), bats (who roost in thin slotted spaces), and birds (who nest on shelf-like spaces and cavities).

Acknowledging that birds and bees perceive the world differently—sensing ultraviolet light invisible to the human eye—the habitat boxes were painted with both white and UV-reflective paint. Under black light, this hidden layer mimics their visual spectrum, offering a glimpse into how these creatures might experience their environment.

→ This image presents a close-up view of hand-carved wooden habitat boxes created from salvaged wood pieces glued into blocks and embellished with intricate details.

Collaborative **Processes and Outcomes**

The process of creating a variety of uniquely designed habitats, each with its own place-based narrative related to the castle, invited collaboration from fellow residents. One of the visual artists, Esteban Cabeza de Baca, principally a painter, spent some time in the Civitella woodshop, working on a habitat design. In this collaborative process, Esteban brought his own artistic vision and sensibilities to hand carving a wooden habitat box, while also bearing in mind the dimensions and spatial parameters of solitary bee holes, a key consideration when designing this unique piece.

The process of collecting and listening to animal recordings not only led to a better understanding of Civitella's ecological landscape but also of how these sounds could be experienced, collectively, as a form of music, to elicit emotion. For example, the giant noctule bat's recorded call—a deep wailing cry—sounds extremely mournful. Listening to the sounds of bats, birds, and frogs in the context of an artist residency with musicians and composers led to another collaboration, with music fellow Daniel Bernard Roumain, who composed and recorded a new twenty-minute track titled "Joy in the Woods," incorporating violin, synthesized music, and a selection of wildlife recordings.

← **Joyce Hwang, Shelter of Belonging, Civitella Ranieri, Umbertide, Italy, 2024**

Acknowledging that birds and insects see ultraviolet light, the habitat boxes are each painted with UV-reflective patterning, revealed by black light flashlight, offering a glimpse into how animals see the world differently.

Architecture **as Interspecies Care**

As a site that is rooted in, and responsive to, centuries of cultural history, Civitella Ranieri presents a context of continuous ecological and social change. The giant sequoia trees adjacent to the installation site—imported to the grounds at the start of the twentieth century—have stood as a constant presence since before the residency's inception, bearing witness to the ways that human visitors have occupied and transformed the grounds over the years. Situated on the periphery between the forest and the castle's manicured landscape, they've seen the ways that residents have adapted the environment to enjoy this liminal space. A hint at one of its past stories are two parallel pieces of rope hanging from a nearby tree, marking a spot where there was formerly a makeshift swing. Past residents probably enjoyed this swing, looking in from the margins. While the ground was being prepared for the installation of the Shelter of Belonging, a tree stump in the middle of a small clearing was uncovered. At some point a tree was there, but why was it removed? Was there a desire to create an open space in the forest for social gatherings? Or was it deemed the only way to tend to a dying tree? Within these continuing narratives of adaptation, the Shelter of Belonging offers a modest contribution as a third space that not only amplifies opportunities for rest—for both visitors and hosts—but perhaps also makes visible the processes of life and death that bring meaning to the built environment and our experiences in it.

1. Jakob von Uexküll, *A Foray into the Worlds of Animals and Humans: With a Theory of Meaning*, trans. Joseph D. O'Neil (Minneapolis: University of Minnesota Press, 2010), 53–72.

Reorienting Ancient Practices for Regenerative Futures

Phil Ayres

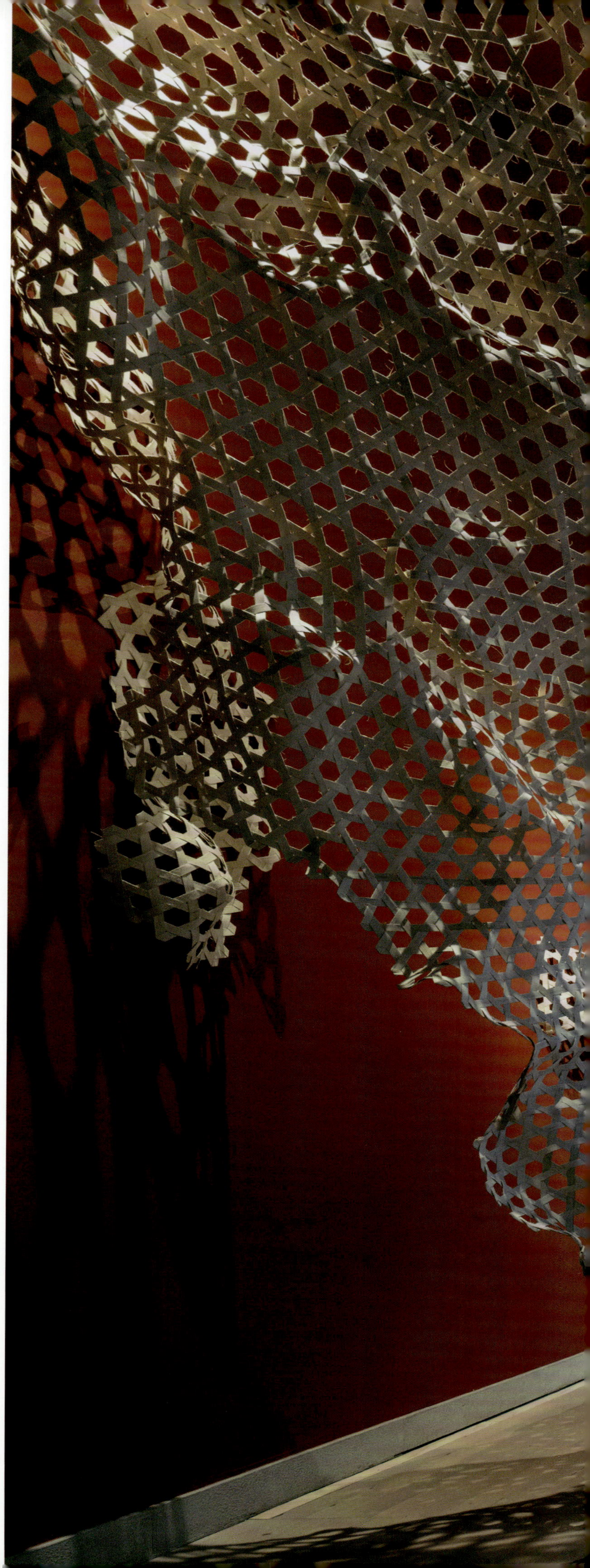

GROTTA DEGLI
SCHIAVI DI SHINOBI

Fermentation and weaving are among the most fundamental practices that have helped shape human development. Fermentation has predominantly served to create and preserve food and beverages; weaving has facilitated the making of clothing, vessels, and shelter. Together, these techniques have been instrumental in underpinning societal developments, resulting in the emergence of modern cultures, economies, and technologies. For example, weaving was a nexus of early industrialization, which saw a craft economy transition rapidly into a fully-fledged factory-based industrialized sector. Furthermore, developments in the mechanization and automation of weaving acted as precursors for modern computation, most directly with the development of technologies for programmable machine control to realize complex weaving patterns. This is perhaps best exemplified by the advent of the Jacquard loom in the early nineteenth century. In the case of fermentation, significant advances in the study of microbial action occurred across the seventeenth to nineteenth centuries through the work of scientists such as Antonie van Leeuwenhoek and Louis Pasteur. However, it was not until Emil Christian Hansen isolated a pure yeast cell at the Carlsberg Laboratory, Copenhagen, in 1883, that fermentation transitioned from an inconsistent artisanal craft into a controlled and predictable industrial process. This not only paved the way for developments in food security but deeply influenced societal health (across the health spectrum), culture, and identity, and acted as a pillar for supporting burgeoning urban and economic growth.

↰ **Cave_bureau in collaboration with Phil Ayres and Jack Young, Shimoni Slave Cave, *GBR—Geology of Britannic Repair* exhibition, British Pavilion, 19th Venice Architecture Biennale, Venice, Italy, 2025**

The Shimoni Slave Cave installation demonstrates how the principles underlying Kagome weaving can be applied to create a large-scale irregular fractal geometry. At this scale of weave, intuition cannot be relied upon to achieve a near-net approximation of the target geometry.

Reorienting Fermentation and Weaving as Catalysts for Transition

The versatility and deep innovative potential of fermentation and weaving have long been articulated and exploited in the interests of industrial development, with associated benefits for many societies (but not necessarily all citizens). However, the limits of an unabated pursuit of this paradigm have also been starkly modeled for at least the last half century. The negative ecological and social impacts are well recognized, largely acknowledged, increasingly felt, and yet they persist. Within this context, and given the versatility of fermentation and weaving techniques, the question arises of whether they might instead be leveraged as the catalysts for alternative courses of action focused on renewal and regeneration.

Fermentation and weaving exist in some form within all human cultures. This suggests that alternative narratives for their use, and novel insights into their application, could, through social diffusion, quickly “scale-out” and “scale-deep” within cultures already receptive and practiced in their use. This could potentially counter the entrenched unidimensional, reductive, and persistent modernist focus on “scaling-up.” Scaling-up in this context generally points to industrial settings, which tend toward the commodification of their outputs, and usually operate within frames that have increasingly reduced mechanisms for the redeployment of generated wealth. However, there is more at stake in terms of expanding a relational field beyond a human social context, since fermentation and weaving are also capacities we share with insects and animals. This carries profound philosophical,

→ **Professor Michael Poulsen, *Acromyrmex* leaf-cutter ants tending *Leucoagaricus gongylophorus*, Section for Ecology and Evolution, Department of Biology, University of Copenhagen, Denmark, 2024**

Fermentation is not an exclusively human practice. *Acromyrmex* leaf-cutter ants cultivate the fungus *Leucoagaricus gongylophorus* to perform fermentation of the leaf matter they collect. The ants are reliant on the metabolic activity of the fungus to convert inedible plant matter into digestible, nutrient-rich structures called gongylidia. This ant colony resides at the University of Copenhagen's Department of Biology.

↓ **Blackbird nest woven into a fragment of Kagome weave, Chair for Biohybrid Architecture, Royal Danish Academy, Copenhagen, Denmark, 2024**

Weaving is not only practiced by humans. This blackbird nest was woven onto a portion of Kagome weave in external covered storage. The bird's weave method contrasts with that of Kagome but achieves double curvature and ridged fixing through a richer palette of materials.

conceptual, and practical significance. For example, existing beyond symbolic language or formal teaching, weaving and fermentation are clearly not exclusive hallmarks of human ingenuity. This challenges prevailing notions of human uniqueness and exclusivity. It also prompts speculation that such techniques may represent evolutionary deep structures—adaptive strategies so fundamental that they predate human history. While weaving and fermentation can lead to sophisticated outcomes, and can be articulated through complex processes, they may by largely dependent on emergent processes. And these emergent processes are governed by simple rules and local interactions that occur across a field of relational agency, material intelligence, and environmental feedback. Approaching them as emergent principles and fundamental ecological behaviors bolsters the idea of employing them as pillars for regenerative design.

The vision pursued by the Chair for Biohybrid Architecture at the Royal Danish Academy in Copenhagen, Denmark, is to develop architectural systems that act as ecological assets rather than as burdens. This is investigated through the development and exploration of structural and material approaches that can integrate and/or couple living complexes. Central to this investigation is the applied use of weaving and fermentation for creating eco-socially contextualized structures, materials, and practices for the built environment. From a social perspective, this research seeks to establish pathways for resocializing and democratizing the practice of architectural design and construction. This is seen as a critical pathway for empowering people to participate in the development of their built environments with the ambition of engendering deeper relations with these environments. The Chair's research insights are therefore intentionally accessible, inclusive, and, where they involve material investigation or processing, not dependent on advanced industrialized technologies in their manufacture. The aim is to foster and support social uptake, social diffusion, translation, and integration into local contexts, with an emphasis on promoting citizen autonomy rather than dependency. The focus on fermentation links directly to bio-based methodologies and biofabrication, while the focus on weaving, specifically a triaxial version of weaving called Kagome, opens perspectives relevant to biology with regard to relationships between topology, geometry, material, and self-organization.

Kagome is a traditional handcraft for making woven baskets. It is a weaving technique that is practiced around the world, although the oldest archeological finds, dating back seven thousand years, suggest it originated in Japan. Indeed, the name Kagome is a Japanese portmanteau derived from two kanji—*kago* and *me*—whose literal translation is "basket eye," in reference to the hexagonal voids that result from interlacing strips of material to follow a trihexagonal tiling. With a vertex configuration $(3.6)^2$ and p6m symmetry, this trihexagonal tiling represents one of the eleven uniform polygon tilings of the Euclidean plane. Following this tiling results in sparse, planar weaves. However, a fundamental property of the Kagome weaving technique is its ability to induce double curvature into the lattice through the introduction of "singularities." A singularity refers to an alternative polygon that has been substituted for a hexagonal cell. Introducing a lesser-sided polygon induces synclastic (positive Gaussian) curvature, leading to elliptical geometry; while the introduction of greater-sided polygons induces anticlastic (negative Gaussian) curvature, leading to hyperbolic geometry.

The more edges that are removed or added to the singularity, the more pronounced the double curvature becomes. This property can be described through the mathematical principles of angular excess and angular deficiency, which are both fundamental concepts for the study of non-Euclidean geometry. These same principles can also help us understand and describe the self-organization of matter in many biotic and abiotic systems where curved space, tessellations, or topological constraints govern structure and function. Examples here include the folding of proteins, the growth of cell sheets and epithelial morphogenesis, plant phyllotaxis, the construction of "imperfect" beehives, crystal growth on

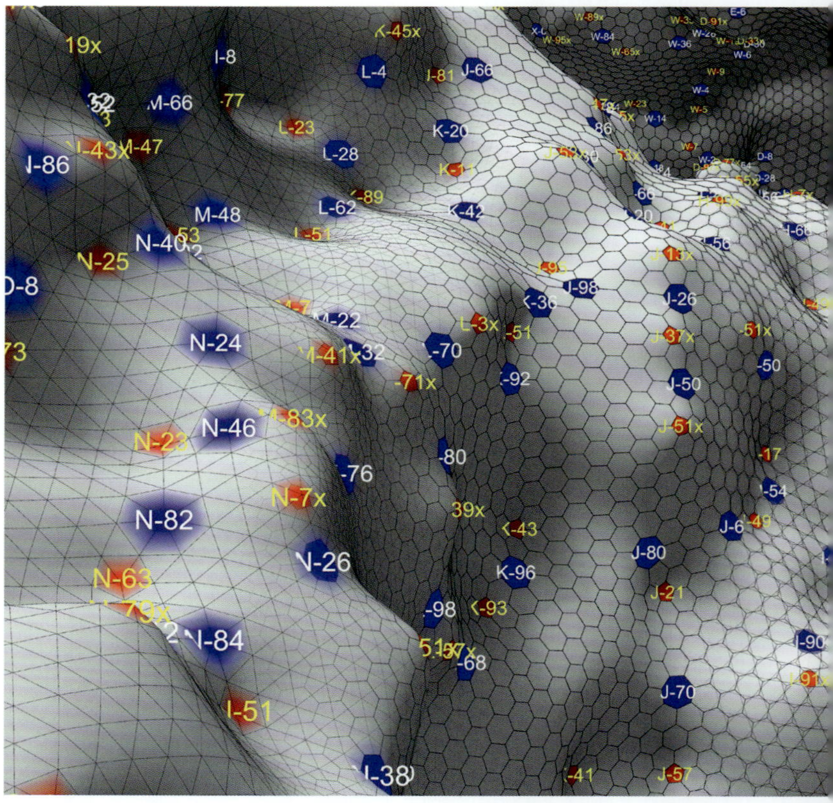

→ **Phil Ayres and Jack Young, stages of weave representation on a generated topology, Kagome Architectures project, Chair for Biohybrid Architecture, Royal Danish Academy, Copenhagen, Denmark, 2024**

Finding a suitable topology for achieving a geometric target can be achieved by creating a surface tessellation of equilateral triangles. Generating the tessellation can be easily accomplished computationally and represents a "topology finding" exercise. Having found a suitable topology, a Kagome weave pattern is made by generating the "in-triangle" of each face. These new triangles produce a principled trihexagonal lattice with embedded singularities that guide physical weaving.

curved substrates, and various forms of geological formation. Kagome therefore acts as a powerful proxy, model, and demonstrator of certain fundamental principles that govern the self-organization of matter. As such, this craft can help foster appreciation and illuminate aspects of the natural world around us, while the principles it embodies can also be applied to make things that extend far beyond the technique's origin in baskets.

In weave design, the ability to locally induce and control double curvature provides a principled basis for achieving near-net approximations of complex morphologies, including high genus and fractal geometries, using only straight strips of material. This also yields highly resilient structures that effortlessly integrate functional utility with potent aesthetic qualities, while the use of locally sourced and abundantly available fibrous materials makes this weaving technique both accessible and inclusive. Furthermore, the production of free-form geometry using straight strips of material provides a rational, systematic, and efficient approach to fabrication. Curved geometry also activates increased structural performance. These attributes make Kagome a very interesting target of study within an architectural context. However, it also requires a shift in design thinking, from a focus on geometry to a focus on topology in interaction with material properties.

Given all of this, the key question from a weave-pattern design perspective is: Where should singularities be located in order to achieve desired morphological outcomes? For many objects, this can be discovered through trial and error, or for those with weaving experience, through the guidance and application of intuition. However, as the geometric

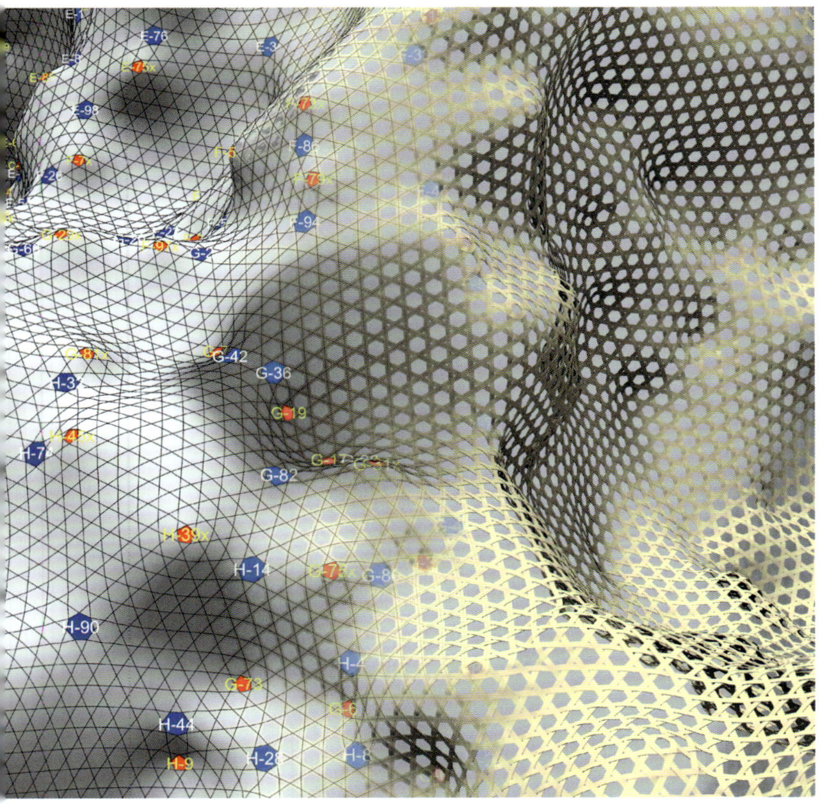

complexity and the number of lattice cells increases, trial and error becomes untenable as an exploration strategy and intuition can be confounded. Irregular fractal geometries are one example where alternative, systematic approaches to developing principled weave patterns, and making fabrication tractable, become essential.

Generating a principled weave pattern to achieve a near-net approximation of a geometrical target is predicated on finding a suitable topology given the cross-sectional size and bending stiffness of the weave material. The methods developed within the Chair for Biohybrid Architecture allow for full automation, full user definition, or a hybrid approach, to cater for different design goals and constraints.

These approaches are scalable (up, out, and deep), as has been demonstrated in the establishing of communities of weavers to produce large-scale structures with complex fractal geometries. Most importantly, the relatively simple computational design pipeline supports the capacity to physically realize geometric complexity through a method that can be taught in a single day. The experience of running these workshops with people from all sorts of backgrounds and of all ages has affirmed that the accessibility and inclusivity of Kagome weaving is empowering. It provides immediate avenues for participating in, and contributing to, the making of a built environment.

The same is true of working with fermentation, through which mycelium-based materials (MBM) can be made. MBM exhibit many physical properties that make them suited for application within construction. They are lightweight, fast growing, can be produced with low embodied energy, and can act as a carbon sink if produced using carefully considered production protocols. Within a building construction context, they can perform well both thermally and acoustically. In addition to these physical properties, MBM production can leverage a variety of existing, readily available, and abundant waste streams. These include residues from the agricultural, forestry, and food sectors. In addition, and in principle, when MBM reach end of life they can be recycled or biodegraded. From a design perspective, MBM provide accessible and rapid production of materials, the properties of which can be tuned toward different mechanical properties, functional poles, and aesthetic characteristics. Four primary parameters provide routes for dialing-in toward material objectives: the choice of fungal strain, the composition of the growth medium, the conditions of cultivation (growth protocol), and post-processing of the resulting material. These create a vast space of exploration and opportunity for investigation within conventional and emerging fields of material development.

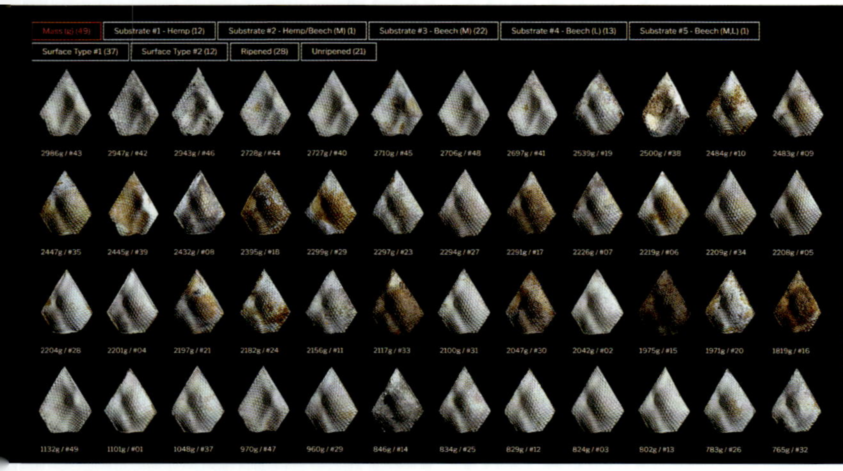

↑ **Phil Ayres, Tereza Vesela, You-Wen Ji, and Adrien Rigobello, mycelium panel production and assembly, Fungal Architectures project, Chair for Biohybrid Architecture, Royal Danish Academy, Copenhagen, Denmark, 2024**

A collection of mycelium-based panels produced using *Ganoderma lucidum* together with various combinations of locally available waste streams and wood particles, resulting in different classes of expression, texture, and mass. The panel geometry is designed to produce a nonuniform tiling with poche spaces that can be articulated toward spatial objectives.

← **Phil Ayres, Jack Young, Claudia Colmo, You-Wen Ji, and Adrien Rigobello, full-scale prototype for a woven and fermented architecture, Fungal Architectures, Chair for Biohybrid Architecture, Royal Danish Academy, Copenhagen, Denmark, 2023**

A physical fragment developed from the design speculation for a woven and fermented architecture. Two scales of weave intersect and perform roles of primary structure and enclosure, respectively. The separation allows for each class of weave to adopt its own geometry, while their intersection generates poche subspaces for undetermined uses and habitations.

Final Endnote
The Kagome Architectures—Novel Woven Construction in Solid Timber project (2024–29) is funded by the Carlsberg Foundation under the Semper Ardens Accomplish instrument, grant number CF23-1255. The Fungal Architectures project (2019–23) was funded by the European Union's Horizon 2020 research and innovation programme, FET Open, under grant agreement No. 858132.

In combination, weaving and fermentation open up fascinating and under-investigated spatial, material, tectonic, and social territories for design inquiry. They also engage wider communities beyond the human, allowing for conscious relations to be established across flora, fauna, funga (FFF), and the broader microbial world. Inviting the participation of these extended communities can engender new relations, even new design paradigms, that exist beyond conventional design boundaries and conventional architectural functions.

A reorientation and repositioning of weaving and fermentation as inclusive methods to invite accessible and informed citizen participation can foster a re-engagement in the making of things and the built environment. Lowering barriers to entry does not equate to a reduction of complexity in outcome, rather, it points to pathways for fostering alternative ways of "being-and-doing" that can align with principles of restoration, regeneration, and social innovation.

→ **Annegrete Mølhave, Silje Eøry Sollien, Madison Lindsay, Phil Ayres, and workshop participants, community-built composting and gifting weave, Ancient Practices for Uncertain Futures workshop, Royal Danish Academy, Copenhagen, Denmark, October 8, 2025**

A Kagome weave acting as a geotextile for framing composting and for the gifting of nesting materials for local birds. The making of the weave and the compost (a related form of fermentation) engages wider communities beyond the human, allowing for conscious relations to be established across flora, fauna, funga (FFF), and the broader microbial world.

↓ **Phil Ayres and Claudia Colmo, design speculation for a woven and fermented architecture, Fungal Architectures, Chair for Biohybrid Architecture, Royal Danish Academy, Copenhagen, Denmark, 2023**

The proposition consciously articulates an indeterminate character poised between building and landscape, decay and renewal, lattice and infill.

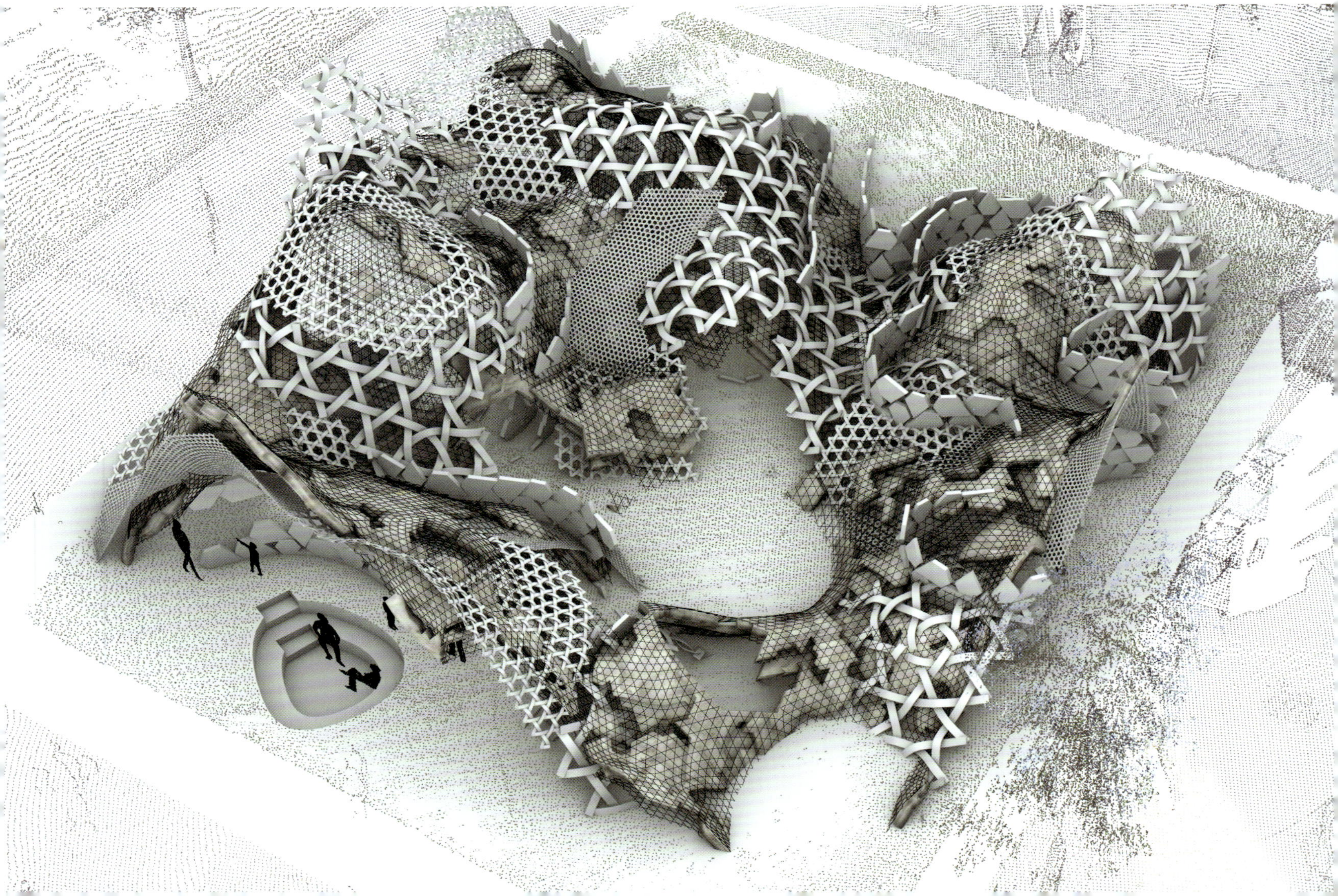

The “Bios” in Bio design

Ehab Sayed

The word "biodesign" is often understood, through the lens of biology, as a scientific discipline. Yet its deeper essence and indeed the origins of the scientific discipline itself, rooted in the Greek word *bios*—encompassing physical, chemical, biological, anthropological, spiritual, and cosmological life—invites us into a much richer terrain. Rather than denoting the sum of mechanical biology, bios constitutes life itself; the cosmos experienced as a complex web of dynamic vitality and cross-systemic relationality. As such, bios-driven design centers reverence and interconnectedness, not control.

To genuinely embrace this vitality, architecture must transcend aesthetic mimicry and instead enter the domain of deep emulation—of processes, systemic relations, and principles honed over 3.8 billion years of evolution. This is the invitation of biomimicry, redefined as "the holistic and systemic emulation of the multitudinous hierarchical systems in nature (the whole universe) to create intelligent solutions that are glocally situated, socially connected, and sensitive to the cybernetics and energetics of their environment."[1] This forces us to shift from an aesthetic emulation that sees nature as a resource or a collection of surface-level mechanisms, to a worldview that sees nature as the place, the source, and the result of material phenomena—and this includes humans, human culture, and human artifacts.

What is Nature?

↰ **Biohm, Terraphyll™ insulation materials, Biohm HQ, London, 2025**

Terraphyll™ panels are produced from agricultural and industrial byproducts using bio-based processes. Their natural composition and precision engineering deliver high-performance thermal insulation while enabling scalable, regenerative construction.

→ **Biohm, Obscure lampshades crafted from Orb, at Apricity restaurant, designed by Object Space Place, London, 2022**

Obscure lampshades embody a circular, nature-inspired approach to material innovation, composed of orange peel or coffee chaff bound with a plant-based binder. Each lampshade diverts organic waste from landfill, functioning as a mini carbon sink, and can be fully composted.

→ **Biohm, waste used in the creation of Orb, London, 2022**

Technicians sort organic waste streams—the first step in transforming discarded materials into regenerative products. By redefining waste as a resource, the process underpins circular production models that are central to biomimetic design.

→ **Biohm, soil-based Orb tiles at The Arbour by Boehm Lynas and GS8, London, 2021**

This version of Biohm's Orb material transformed topsoil into lampshades, shelving, and headboards at The Arbour, a carbon-negative development on a former industrial site in Walthamstow Village, East London, in collaboration with GS8.

Western constructions of nature have historically rested on binaries[2]: nature and culture, natural and artificial, human and nonhuman. Such divisions, deriving from Ancient Greek dualism and subsequently enhanced by Enlightenment rationalism and Judeo-Christian thought, engender a worldview oriented toward domination and extraction. Nature thus becomes passive, inert, and "out there."

Non-Western cosmologies, however, remind us that such attitudes have not always been the case. For the Taoist, nature is a flowing totality of which humans are an indivisible expression. For the animist,[3] nature is not a collection of objects but a communion of subjects. In the Shinto-infused techno-animism of Japan,[4] even the artificial is enfolded into nature—robots, machines, and animated spirits are alive in their own way, animated by intention and integrated into ecological relationships. The animator Hayao Miyazaki's worldbuilding perfectly embodies the meshing of artificial and natural, depicting forests and factories not as opposites but as interpenetrating forces.[5] These worldviews reveal that seemingly entrenched categories are contingent, cultural, and ripe for renewal.

This philosophical reframing enlivens a more expansive understanding of biomimicry—catalyzing a design method that also cultivates cultural and ontological shifts. It transforms buildings from inert monuments into evolving metabolisms, blurring the line between the natural and the artificial. Indeed, planetary anthropogenic manipulation renders the persistence of any truly "natural" phenomena uncertain.

Biomimetic Intelligence in Practice

This is where biomimetic intelligence becomes essential. Defined by three interlocking priorities—systemic integration, adaptive capacity, and regenerative potential—it signals a paradigm shift from mechanistic problem-solving toward evolutionary participation.[6] It calls on architects and designers to embody the strategies evolved by systems in nature to meet similar challenges: scarcity of resources, climate fluctuations, structural integrity, energy flows, and interdependence. Unlike human technologies, which often prioritize speed, scale, and singular functions, biological systems optimize for multifunctionality, cooperation, and sufficiency. Biological solutions often sharply diverge from human-engineered approaches as nature tends to favor hierarchy, feedback loops, and information flows over brute force and material excess.[7]

A compelling manifestation of this thinking emerged in the research collaboration between Biohm and the property company British Land (2021–22), which sought to map the hidden metabolisms of an urban campus by tracing pathways of operational and construction waste. Biomimetic strategies revealed waste streams that were not "waste" at all but rather nutrients misplaced by misaligned systems. With systemic analysis, stakeholder mapping, and scenario modeling, a new picture emerged: one where closed-loop material flows, localized production, decarbonization, and regenerative outputs were not only possible but economically prosperous. This resulted in the creation of Organic Refuse Biocompound (Orb), a sustainable construction material made from waste byproducts from the agricultural and food production sectors, which can be used in sheet form, as well as molded, compressed, or 3D printed.

The principles guiding this collaborative study also informed a range of circular resource management projects in London, including acoustic paneling and Orb at County Hall (2023), Orb wall tiles and lampshades at The Arbour residential development (2021), and Orb lampshades found at the restaurant Apricity (2022). Biomimetic intelligence alters the logic of sustainability and expands the concept beyond our preconceived anthropocentric notions. The objective shifts from harm reduction to the pursuit of a biosynergetic future where architecture is an act of ecological generosity, returning more to the environment than it extracts.

A Deeper Biomimicry Matters

True bio(s)design, rooted in a grounded biomimetic philosophy, requires the extension of this regenerative approach beyond the design of buildings and into the design of components and products.[8] This invitation toward a deeper biomimicry is exemplified in Terraphyll™, Biohm's building insulation technology, which synthesizes a plethora of complex challenges into a single, simple solution. The Terraphyll™ manufacturing system is being developed to emulate nature's own creation of matter: breaking down complex substances into simple building blocks that can be bioengineered and reassembled for a new life.

Terraphyll™ revolutionizes the industrial use of fungi in continuous scaled manufacturing by converting a wide range of engineered lignocellulosic agricultural and industrial waste—including corn, soy, hemp, straw, timber, paper, and board—into high-performing materials with consistent performance properties. This process disrupts traditional supply chains by eliminating dependence on finite raw materials, replacing them with abundant, decentralized, and hyperlocalized feedstocks. It is the first bio-based material that does not rely on a specific feedstock and is therefore both globally scalable and highly resilient in the face of rising geopolitical uncertainty.

Terraphyll's™ precise control over material properties, despite its varied inputs, results in a regenerative material that meets construction industry standards. In fact, Terraphyll™ rivals the leading synthetic insulation, with top-tier performance properties (thermal conductivity: <0.036 W/mK) and price points that undercut established alternatives. But more importantly, it is carbon negative—sequestering 44 kg CO_2e per m^3—thereby

← **Biohm, mycelium acoustic panels and Orb material boards, at County Hall, London, workspaces designed by Material Works, 2023**

Biohm's Orb material boards, made from agricultural waste, and mycelium acoustic panels used in the Sustainable Workspaces in County Hall, London.

← **Biohm, Terraphyll™ insulation, Biohm HQ, London, 2025**

Biohm's Terraphyll™ insulation development process consists of rigorous evaluation to ensure the material meets stringent building standards while advancing regenerative construction solutions.

transforming buildings insulated with Terraphyll™ into carbon sinks. Additionally, testing by the British Board of Agrément, the construction industry approvals body, gave the product an A+ indoor air quality rating and found that it released zero toxic chemicals, making it one of the healthiest insulation materials tested. Once scaled, it will liberate manufacturers from highly controlled and volatile chemical markets, replacing them with increased vertical integration, local inputs, and the gift of self-replicating fungi—and yielding profit margins of 30–70 percent compared to the industry's typical <10 percent. High profit margins can facilitate the integration of social regeneration during production by embedding profit-share schemes for local communities within one hundred kilometers of each factory.

Holistic solutions, rooted in a deeper biomimicry, enable us to build nodes of localized regeneration around the globe in a way that does not contradict nor create conflict with current industrial, economic, and business needs. When the complexity of nature is built into our thinking and decision-making, we can create simple solutions that intelligently meet the needs of the complex systems within which they are situated.

The "bios" in "biodesign" is not a prefix. It is a provocation for an architecture that lives and participates in nature's primordial intelligence. It means becoming, once again, a humble part of the whole, and seeing that the line between nature and architecture was always a fabrication.

1 Ehab Sayed, "Biomimetic Intelligence for Systemic Innovation in the Built Environment" (PhD diss., Northumbria University, Newcastle, 2022).

2 René Descartes, *Le monde de Mr. Descartes, ou Le traité de la lumière et des autres principaux objets des sens: avec un discours de l'action des corps et un autre des fièvres, composez selon les principes du même auteur* (Paris: Girard, 1664).

3 John R. Clammer, "The Politics of Animism," in *Figured Worlds: Ontological Obstacles in Intercultural Relations*, ed. John R. Clammer, Eric Schwimmer, and Sylvie Poirier (Toronto: University of Toronto Press, 2004), 83–109.

4 Casper B. Jensen and Anders Blok, "Techno-Animism in Japan: Shinto Cosmograms, Actor-Network Theory, and the Enabling Powers of Non-Human Agencies," *Theory, Culture & Society* 30, no. 2 (2013): 84–115.

5 Shoko Yoneyama, "Miyazaki Hayao's Animism and the Anthropocene," *Theory, Culture & Society* 38, nos. 7–8 (2021): 251–66.

6 William E. Odum, Eugene P. Odum, and Howard T. Odum, "Nature's Pulsing Paradigm," *Estuaries* 18, no. 4 (1995): 547–55.

7 Julian F. V. Vincent, Olga A. Bogatyreva, Nikolaj R. Bogatyrev, Adrian Bowyer, and Anja-Karina Pahl, "Biomimetics: Its Practice and Theory," *Journal of The Royal Society Interface* 3, no. 9 (2006): 471–82.

8 Ehab Sayed, "Offsite Manufacturing Innovation: A Biomimetic Future," in *Offsite Production and Manufacturing for Innovative Construction*, ed. Jack S. Goulding and Farzad Pour Rahimian (London: Routledge, 2019), 419–48.

Bioma

Mae-ling Lokko

Imagi

terial

naries

Future Biomaterial Imaginaries

Societies first and foremost establish themselves through their imaginaries. These imaginaries become the forges from which the forms that a society grants itself can emanate for nourishing life as well as enriching it, for elevating human creativity and the adventure of social invention to a higher level. These imaginaries also evolve because they are projected into the future; they think the conditions of their own longevity.

Felwine Sarr, *Afrotopia*[1]

In an era filled with ecological anxiety and economic inequity, the umbrella of biodesign has reinstated plant materials and living systems to their rightful place at the contemporary design table, destigmatizing their use and reigniting the belief that humans can work *with* and not *against* nature to build, inhabit, and maintain buildings.

Over the last two decades, the use of plant remains—husks, stalks, shells, sheaths, rinds, leaves, peels, and stems—has emerged as a powerful strategy to address long-standing inequities in global material value chains, particularly in regions historically shaped by extractive agricultural economies. This change began by engaging the vast surplus generated by African agricultural industries—sectors that have long exported low-value crops while facing barriers to accessing affordable, contemporary construction materials. More recently, this approach has opened new imaginaries around the architect's role in generating, translating, and circulating value across twenty-first-century academic, industrial, and community contexts.

Inspired by "generative justice" systems, this work has led to the development of intersectoral enterprises and collaborations as integral components of design research.[2] One such initiative is AMBIS Technologies Inc., a building material startup launched in 2015 to connect agricultural byproduct processors with emerging bioadhesive companies in post-manufacturing industries.[3] Supported by a startup accelerator and an interdisciplinary research network, AMBIS has served as a model for embedding biomaterial research within academic contexts while exploring its architectural applications across diverse markets.

Given the sheer scale of bio-based material demand required to meet half the global housing needs by the end of the century, biodesign has also opened new economic imaginaries for actors that have typically been undervalued or invisible across sectoral material life cycles.[4] Initiatives like AMBIS have facilitated collaborations with small-scale farming cooperatives and waste collection enterprises, engaging actors at both ends of the plant material life cycle. For these partners, rising demand for agricultural residues, forestry byproducts, and invasive species presents new opportunities within a green-collar economy—capturing value that would otherwise be prematurely downcycled, burned, or discarded.

← **Mae-ling Lokko, Mycotunnel, Z33 House of Contemporary Art, Design and Architecture, Belgium, 2022**

Located within a two-mile radius of RIBA North gallery, a series of grow-it-yourself (GIY) workshops were staged at Squash Nutrition, a creative food enterprise in Liverpool, UK. The exhibition *Hack the Root* explored distributed infrastructures and networks for growing bio-based materials. On-site growing, production, and installation of the mycelium "tunnel" structure was central to the exhibit experience. The Mycotunnel installation from *Hack the Root* was rebuilt four years later for a solo exhibition at Z33 House for Contemporary Art in Hasselt, Belgium.

Despite clear demand for products like insulation, medium-density fiberboards, and acoustic panels in domestic markets across the global majority, two key barriers persist: higher costs compared to conventional materials and entrenched negative perceptions around their use. To address these challenges, Willow Technologies—a biomaterials startup based in Accra, Ghana—was established in 2018 to focus on accelerating the diffusion of bio-based materials and expanding access to their distributed production across diverse communities.

Architectural and art exhibitions have served as critical platforms for reimagining the cultural identities of biomaterials and for scaling experimental products at key moments in the design research process. *Hack the Root*, part of the 2018 Liverpool Biennial, exemplified this approach by demonstrating how mycocomposites could be produced through mycelium workshops using simple grow-kits, with materials cultivated on-site in the gallery. Over the course of three weeks, more than two hundred participants from K–12, university, and urban farm environments joined these workshops throughout the city, highlighting the potential for public coproduction and decentralized fabrication.

In recent years, exhibition-based design research has increasingly focused on symbolic spaces of extraction as critical sites for rethinking the aesthetic and cultural paradigms of biomaterials—from their physical performance to the narratives they embody. These installations have explored themes such as human health, biophilia, and probiotic imaginaries, which have become key environmental and social drivers for the adoption of biomaterials.[5] Architectural settings have included the mining tunnel, the "door of no return" found in West African slave forts, the plantation, dining table, and the botanical greenhouse. Across these contexts, biomaterial systems have consistently applied modular, human-scale components—typically panels of 1–1.5 square foot—designed for low-energy growing and casting, rather than for thermal pressing. Mold designs and production energy demands have gradually evolved to align mechanical, hygrothermal, customization, and design-for-disassembly logics with distributed, nontoxic, and low-carbon production models.

New Paradigms for Biodiversity Across the Building Life Cycle

The whole life carbon story of bio-based building materials has been envisioned as a triple-heavyweight for atmospheric decarbonization—plants sequester carbon as they grow while bio-based building materials delay carbon emissions—storing and reducing the operational emissions through effective hygrothermal design and integration. Yet to capture such decarbonization benefits requires both a direct and indirect expansion of biodiverse resources to meet material performance standards and thermal comfort goals.

Under a capitalist system, scaling bio-based materials for building sectoral demand can risk extraction rates that far outpace ecological regeneration.[6] More critically, the development and marketing of biomaterials as mere substitutes for their high-carbon mineral incumbents, without a parallel shift in antibiotic indoor environmental cultures, simply displaces toxicity further down the building life cycle and maintains the building sector's leading role in driving higher carbon emissions.

As the "big brother" leading the bio-based materials sector, timber and engineered wood production has paved the way for scaling global supply chains from living systems, modeling the "buildings as carbon sink" paradigm.[7] By targeting the use of high-strength biomaterials to replace the high-mass and high-carbon structural components of contemporary construction systems, the use of thermally pressed and laminated biomass technologies has revealed the entrenched monoculture plantation at the root of carbon sink products. Today, the mass timber industry primarily relies on a limited range of two to three tree species to meet demand for standardized products.

Similarly, the greatest potential for using plant residues, additives, substrates, and binders in bio-based building materials relies on super crops like sugar, corn, rice, wheat, and soy.[8] Despite widespread experimentation with scaling mycocomposite bio-based materials technologies, commercialization in the building sector has largely depended on just a few mycelium strains.[9] Moving toward biodiversity goals—by expanding the design of bio-based materials to match seasonal patterns and plant diversity within the bioregion—opens up new opportunities for intersectoral partnerships across forestry, ecomanufacturing, and built environment actors.

To address the limitations of relying solely on coconut husk byproducts from monoculture plantations, for example, the development of AMBIS fiberboard panels expanded to include other plants commonly grown on smallholder coconut tree farms. Corn husk–fed mycelium formed the bulk of the panel mass, while mechanically robust coconut fibers, bound with soy protein, were used for the exterior layer.[10] This layered approach resulted in a lighter panel with better thermal and acoustic insulation properties.

↗ **Mae-ling Lokko, Threshold of Return, Z33 House for Contemporary Art, Design and Architecture, Hasselt, Belgium, 2022**

This installation made of coconut husk panels reconstructs the iconic Door(s) of No Return—symbolizing the extraction of commodities and the departure of human bodies from West Africa—as a site of return. In contrast to the vista of the Atlantic Ocean, this doorway frames a video projection that instead looks reflexively back at the West African coastal landscape.

↘ **CASE-RPI, AMBIS coconut fiberboard panel prototype, Troy, New York, 2016**

Developed in collaboration with Ecofibers Gh. Ltd. and the New York–based companies Ecovative and e2e Materials, this biocomposite panel integrates coconut husk fibers with mycelium and soy-based binders. The prototype emerged from doctoral research (Lokko, 2012–16) into sustainable material systems.

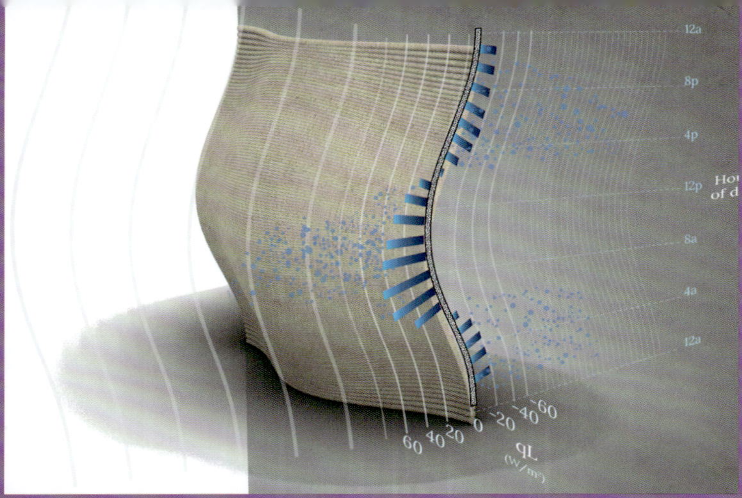

↑ **Mae-ling Lokko, diurnal intrinsic evaporative cooling patterns across a biocomposite panel, Accra, Ghana, 2016**

This panel demonstrates the humidity buffering and evaporative cooling behavior of coconut fiber-based biocomposites in a hot-humid climate. Developed through cross-continental collaboration during doctoral research (Lokko, 2012–16), the material reflects both ecological performance and local resource integration.

→ **Mae-ling Lokko and Gustavo Crembil, Vena Cava, British Pavilion, 19th Venice Architecture Biennale, Venice, Italy, 2025**

From Vena Cava's timber frame were hung patterned panels of fly ash, biomass, bioplastics, and fungi—emerging twenty-first century material streams that herald a shift away from thin mineral-based building envelopes dependent on mechanical environmental regulation, which the Palm House greenhouse at Kew Gardens, London, pioneered in 1848.

Against Botox Culture: Designing Biomaterials for Sweat and Ageing

Often, today's bio-based material success stories reveal conflicts, incongruencies, and standards inherited from contemporary material cycles. Durability and mechanical strength have come at the cost of higher carbon footprints and the increased use of toxic chemicals. A recent study showing the carbon footprint of laminated bamboo products as equal to that of steel, for instance, exemplifies how the sustained use of glues, coating, and additives similar to those used in laminated timber has led to a tripling of carbon emissions and toxicity.[11] In opposition to such "botox" approaches intended to improve the near-term strength, moisture, fire, and durability of bio-based products, the recovery of long-standing bioclimatic traditions of designing breathable materials can ensure the nontoxic circularity and ultimate degradation of biomass materials.

Across the world, comfort has historically been achieved in varied climatic contexts through an understanding of soil and plant wall thickness, density, and their cyclic hygrothermal behaviors over time. From the thick adobe walls of stepped Pueblo houses in arid desert climates to the porous, plant-fiber walls of the tropical Malayan *kampung*, humans have historically wrapped themselves and their houses with a diversity of sweaty soil and plant materials to defend against climatic forces. Historically, the craft of bundling, layering, weaving, and thatching into thicker sections, with composite thermal and maintenance benefits, compensated for lower plant material strength. However, building materials today are often developed and chosen according to siloed performance criteria.

In as much as the prevention of bodily perspiration has been the subject of design since the 1900s, a parallel shift to sealed building envelope systems and mechanical conditioning technologies has occurred. Developed in 2016 with e2e Materials, a soy-bioadhesive company, AMBIS designed building panels made from coconut fiberboards pressed to medium densities to improve their hygrothermal performance. The resultant coconut panels absorbed significant indoor moisture during humid night-time and early morning hours. Over the course of the next day, exposure to high solar radiation and drier daytime conditions led to evaporation of the sorbed moisture, providing a pronounced cooling effect termed "intrinsic evaporative cooling."[12] Through the valuing of porosity as a desirable material property, a broad range of both plant and fungi materials can offer hygrothermal adaptability.

Centering Soil Health as a Goal for Integrated Material Life Cycles

In 2019, two textile waste projects highlighted a critical gap in sustainable design discourse: While the field has often prioritized reducing carbon emissions and material toxicity, it has largely overlooked the impacts of these materials on soil ecosystems. Given the degradation of twenty-first-century soil systems, prioritizing soil health in material design expands the focus from sourcing to a broader intersectoral effort—planning for how materials *return to and regenerate the soil*.[13] This exploration led to the development of "Soil Sisters," an ongoing research initiative at the Yale Center for Ecosystems in Architecture, in New Haven, Connecticut. Soil Sisters encompasses a range of biocomposite material classes that support both immediate relief for soil ecosystems and long-term regenerative impacts.

Engaging twenty-first-century secondary material streams, building materials largely composed of bio-based substrates, nontoxic binders, and additives can play a role in diluting and safely storing a broad range of byproducts from the plastic, textile, paper, glass, and metal sectors. Built in 2025, two architectural pavilions, Vena Cava at the Venice Architecture Biennale and the Ileum pavilion at the Hause der Kulturen der Welt (HKW) in Berlin, took on large-scale botanical greenhouse typologies as sites for Soil Sister material integration. In both, byproducts from twenty-first-century enterprises including agriculture (rice husk, fly ash), metal manufacturing (ferro-silicon alloy byproducts), textile waste, and renewable energy systems (recycled solar panel glass fibers), were employed as substrate, binder, and additive in earthen composite panels using low carbon material life cycle processing.

Although "building as material bank" approaches are needed to address the near-term urgency of environmental loads, the ultimate imaginaries around bio-based materials aim to recouple the matching of building products with regenerative farming practices that underpin the health of soil ecosystems. Relative to short-lived contemporary material paradigms where value generation begins at extraction, is multiplied through transformation, and then steadily decreases throughout its first life, an integrated life cycle approach can drive new metrics along which value is created at strategic points across multiple material lives and actors. Ultimately, value is measured through the proportional use, life cycle timing, and whole life cycle contribution of such materials to the health of the soil, a primary site for sustaining biotic life.

← **Mae-ling Lokko and Gustavo Crembil, Ileum Pavilion exterior view, HKW Pavilion, Berlin, Germany, June 2025**

The 2025 Ileum Summer Pavilion at the Haus der Kulturen der Welt in Berlin drew upon the city's Großes Tropenhaus (a large tropical greenhouse, located in the Botanischer Garten in Dahlem) as a reference site for repair of both soil ecosystems and contemporary building cultures.

→ **Mae-ling Lokko and Gustavo Crembil, Ileum Pavilion interior view, HKW Pavilion, Berlin, Germany, June 2025**

Developed in collaboration with three material manufacturers, Opovate Ltd. (mycelium), Ricehouse (rice husk and geopolymers), and the OR Foundation (textile waste), the pavilion's paneling system was made with composites including biogenic materials and a growing scale of shredded textile cotton waste and secondary metal and solar panel industry byproducts.

1 Felwine Sarr, *Afrotopia*, trans. Drew S. Burk and Sarah Jones-Boardman (Minneapolis: University of Minnesota Press, 2020), 10.

2 Mae-ling Lokko and Ron Eglash, "Transforming the Poor Man's Building Block: Value Creation, Translation and Circulation for Upcycled Indigenous Building Materials," *Folio Journal of African Architecture* 1 (2017): 8–12.

3 Mae-ling Lokko, "Coconut Diaries," *e-flux Architecture*, June 25, 2021, https://www.e-flux.com/architecture/survivance/382526/coconut-diaries.

4 UNEP and Yale Center for Ecosystems in Architecture, *Building Materials and the Climate: Constructing a New Future* (Nairobi: United Nations Environment Programme, 2023), 52–65.

5 Vivian Loftness and Megan Snyder, "Sustainable and Healthy Built Environment," in *Sustainable Built Environments, Encyclopedia of Sustainability Science and Technology Series*, 2nd edition, ed. Vivian Loftness (New York: Springer Science + Business Media, 2020), 87–114; Terrapin Bright Green, *The Economics of Biophilia* (New York: Terrapin Bright Green, 2012), 4–28.

6 Qian Zhang, Yumeng Li, Chang Yu, et al., "Global Timber Harvest Footprints of Nations and Virtual Timber Trade Flows," *Journal of Cleaner Production* 250 (2020): 2–9; Florence Pendrill, U. Martin Persson, Javier Godar, et al., "Agricultural and Forestry Trade Drives Large Share of Tropical Deforestation Emissions," *Global Environmental Change* 56 (May 2019): 1–10.

7 Galina Churkina, Alan Organschi, Christopher P. O. Reyer, et al., "Buildings as a Global Carbon Sink," *Nature Sustainability* 3, no. 4 (2020): 269–76.

8 "Food and Agriculture Data," Food and Agriculture Organization of the United Nations, accessed January 20, 2025, https://www.fao.org/faostat/en/.

9 Alemu Digafe, Mesfin Tafesse, and Ajoy Kanti Mondal, "Mycelium-Based Composite: The Future Sustainable Biomaterial," *International Journal of Biomaterials* 2022, no. 1 (2022): 3.

10 Mae-ling Lokko, Anna Dyson, and Alexandra R. Rempel, "Development of Affordable Building Materials Using Agricultural Waste By-Products and Emerging Pith, Soy and Mycelium Biobinders," in *32nd International Conference on Passive and Low Energy Architecture Proceedings*, ed. Pablo La Roche and Marc Schiler (Los Angeles, 2016), 881–87.

11 Xiaoxiao Xu, Peiyu Xu, Jianjun Zhu, Haitao Li, and Zhenhua Xiong, "Bamboo Construction Materials: Carbon Storage and Potential to Reduce Associated CO_2 Emissions," *Science of the Total Environment* 814 (March 2022): 11–12.

12 Mae-ling Lokko and Alexandra Rempel, "Intrinsic Evaporative Cooling and Weather-Responsive Natural Ventilation for Adaptive Thermal Comfort in Tropical Buildings," *International Building Physics Conference 2018*, Syracuse, New York, September 23–26, 2018.

13 Mae-ling Lokko, "Soil Sisters: Emerging Intersectoral Material Designs For Soil Health," *Journal of Architectural Education* 79, no. 1 (2025): 154–66.

Claudia Pasquero and Marco Poletto

The Soft Technology *of Nature*

ecoLogicStudio's Deep.Forest Transforms Forest Processes into Immersive Biodigital Architecture

↰ **ecoLogicStudio, Deep.Forest, *Living Structures* exhibition, Louisiana Museum of Modern Art, Humlebæk, Denmark, 2024**

View from the entrance of the Deep.Forest immersive exhibition. In the foreground, the system of aeration pipes leading from the ceiling into the engraved cork flooring, inspired in its morphology by the growth of mycelium, guided the visitors along suggested paths.

↑ Deep.Forest focused on the symbiosis of mycelium with photosynthetic microalgae. 102 carbon storers made of salvaged birch trunks hosted a variety of lichen species and provided context for the photosynthesizers—the forty-four glass vessels hosting living cultures of cyanobacteria and seaweeds that supplied the exhibition space with fresh oxygen and carbon sequestration.

→ Axonometric view of the installation.

As we navigate the intertwined crises of climate change, biodiversity loss, and resource depletion, the discipline of architecture is facing a profound reckoning. The limits of conventional design are becoming increasingly apparent. Rigid formalisms, centralized infrastructures, and zoned systems are no longer supporting the complex ecologies of contemporary life. What we need is not merely sustainable architecture but *living* architecture: systems that breathe, metabolize, regenerate, and adapt.

ecoLogicStudio's Deep.Forest installation emerges from this imperative as both a speculative vision and a tangible design framework. Conceived as a cybernetic forest, it proposes a hybrid ecosystem where artificial intelligence, biological systems, and human agency coexist and collaborate. At its core lies a commitment to pursue new frontiers in biodesign through the integration of artificial and biological intelligence into the fabric of buildings and cities.

From photosynthetic air-purifying furniture in London offices to algae-infused biopolymers and mycelium-grown structural elements, ecoLogicStudio's work explores the untapped potential of nature as a soft technology. Programmable, responsive, and regenerative, Deep.Forest marks a new chapter in this journey: a multisensory forest organism that invites the public to enter a world where nature and computation are no longer at odds, where the self is allowed to dissolve in the more-than-human, the Deep dreaming of the Deep.

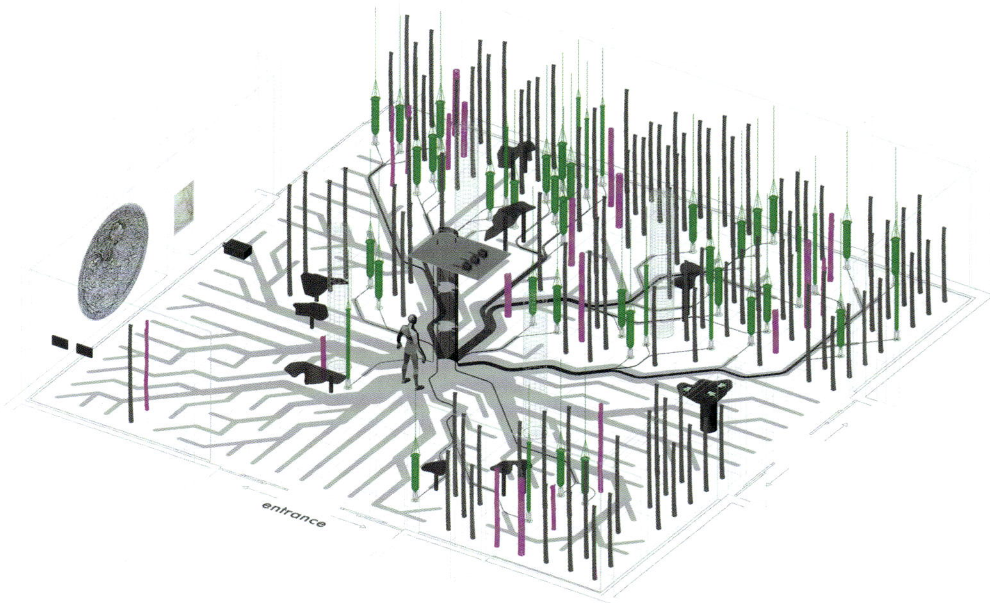

Deep Green

At the heart of the Deep.Forest project lies a radical shift in how we conceive of nature. No longer a static backdrop to human activity, nature is understood not merely as a resource to be preserved but as an active, intelligent, and programmable medium. This is the foundational premise of ecoLogicStudio's biodesign philosophy, a framework that fuses biological processes with computational design to create living, evolving architectures. In this context, architecture becomes not an object but a system, not an endpoint but an interface that operates between species, materials, technologies, and ecologies.

This approach draws from the legacy of cybernetics, biosemiotics, and posthuman thought, proposing that natural systems—like forests, algae colonies, or microbial biofilms—are not only life-supporting environments but adaptive intelligences. These systems sense, process, and respond to external stimuli, much like digital systems do, but with metabolic and regenerative capacities that go far beyond the efficiency of machine logic.

ecoLogicStudio's biodigital practice tested this approach in systems like Air.Garden (2025), developed for the London office of consumer healthcare company, Haleon. This project integrates microalgae *Spirulina* into 3D-printed architectural modules that purify air while producing nutrient-rich biomass. Air.Garden demonstrates how nature's metabolism can be harnessed in design to create circular feedback loops, closing the gap between consumption and regeneration. The byproducts of biological activity (oxygen, biomass, and purified air) become raw materials for fabrication, nourishment, and spatial expression.

Biodesign, in this light, is not merely a new stylistic agenda but an ontological repositioning of architecture itself, from inert object to living medium. It asks us to co-author with nature, to embed responsiveness and reciprocity into the built environment, and to design not just *for* but *with* the planet's multiple intelligences.

It is this ethos that informs the two iterations of Deep.Forest: the original immersive environment at the Louisiana Museum of Modern Art, Denmark, in 2024, and the DeepForest[3] presented at the Triennale Milano International Exhibition, Italy, in 2025. Each explores different spatial, material, and perceptual strategies to engage visitors in the experience of living architecture.

Designing with the **Logics** *of Growth*

The Deep.Forest project stands at the intersection of biological regeneration and material innovation, offering a tangible response to the urgent need for circular, low-impact architectural systems. Across its actualizations, it embraces a vision where waste and pollution are no longer a byproduct but a primary resource.

Algae cultures, specifically *Spirulina* and *Cyanidium*, are cultivated in borosilicate glass bioreactors that not only capture carbon dioxide from the atmosphere but also generate biomass, which is harvested, dried, and converted into biodegradable polymers. These are then used in 3D-printing processes to form architectural skins, frames, and enclosures. In the DeepForest[3] installation, for instance, the printed components mimic tree bark, creating permeable surfaces that support microbial and fungal growth while maintaining structural integrity.

Spent coffee grounds, a ubiquitous form of nutritious, urban, organic waste, became part of another method developed in collaboration with the team of researchers at the Synthetic Landscape Lab at the University of Innsbruck in Austria specifically for the Deep.Forest project (2024). This new method converts coffee grounds from waste into a valuable substrate for growing mycelium. Mycelium forms a fungal network that metabolizes the sugars in the grounds and expands to fill the 3D-printed barks turned architectural formwork. Over time, the mycelium becomes both a structural binder and a living skin, producing a "biodegrader"—an architectural component that processes waste, sequesters carbon, and ultimately becomes part of a new augmented urban biosphere, or "urbansphere."

This augmented biological metabolism is made visible and becomes experiential, even ornamental, so that visitors can observe photosynthesis in action, trace fungal growth, and engage with an architecture that breathes, grows, and decays.

The Deep.Forest design method, in other words, doesn't simply mimic nature. Rather, it performs as a cyber-organic assemblage. Through this regenerative model, Deep.Forest redefines the life cycle of architectural components: from carbon capture to material fabrication, from structural performance to biological decomposition. Each stage is part of a continuous cyber-organic loop, demonstrating that the future of sustainable design lies not in minimizing impact, but in maximizing contribution, to air, to soil, and to society.

↑ **ecoLogicStudio, Deep.Forest, *Living Structures* exhibition, Louisiana Museum of Modern Art, Humlebæk, Denmark, 2024**

Close-up view of one of the twenty biodegraders, 3D printed with algae-infused biopolymer, which contained coffee grounds serving as substrate for colonies of living mycelium. These fed on the sugars in the coffee grounds and grew to form a new space-filling network, a solid synthetic trunk.

→ Plan of Deep.Forest dictated by the satellite imagery processed into an algorithm that mimicked the hidden "Wood Wide Web" of mycelium in a forest.

Deep.Forest, exhibited at the Louisiana Museum of Modern Art, marked the inaugural and foundational iteration of the Deep. Forest series. Commissioned as part of the museum's *Living Structures* exhibition—curated by Mette Marie Kallehauge for the *Architecture Connecting* series—the installation offered a speculative and experiential landscape where architecture became a living, breathing participant in the cycles of carbon, energy, and life.

Rooted in the tradition of site-specific artwork, the installation used satellite imagery from the Copernicus Sentinel-2 network to analyze the vegetation density surrounding the museum. These data were transformed into spatial logic by a bespoke algorithm inspired by mycelial networks, the so-called "Wood Wide Web," creating a layout that mirrors the morphological complexity of a Nordic forest. Within this deconstructed woodland, architecture is reborn as an orchestrator of life processes, an active medium for photosynthesis, biodegradation, and carbon storage.

Upon entry, visitors were immersed in a grid of Deep.Trees, each performing one of three ecological functions: photosynthesizing, biodegrading, and carbon storing. Forty-four photosynthesizers, glass vessels hosting cyanobacteria and local seaweed, generated fresh oxygen and sequestered carbon dioxide from the gallery air. Fifteen of these housed macroalgae from Denmark's coasts, while twenty-nine contained *Spirulina*, a microalgae species capable of producing biomass at rates ten times greater than terrestrial plants.

This harvested algae biomass was then used in a closed-loop process to 3D print twenty biodegraders, bio-architectural elements grown at the Synthetic Landscape Lab at Innsbruck. These structures were made from algae-infused biopolymer mixed with spent coffee grounds, which serve as a substrate for living mycelium colonies. Over a period of weeks, the mycelium metabolized the sugars in the waste, expanding to form a solid structural network that was then cured and dried to become a visually rich, materially resilient component. Occasionally, these forms remained partially wet, allowing mushrooms to fruit from within their bark, reminding visitors of their organic origins and ongoing vitality. Meanwhile, salvaged birch trunks, repurposed as carbon storers, provided habitat for lichen species, drawing a direct line between waste regeneration and biodiversity support.

These elements embody a provocative inversion of the modernist ethos: Instead of mechanizing nature, Deep.Forest naturalizes architecture and computational technology, offering the experience of diving deeply into a cyber-organic forest. It seeks the opportunity of losing oneself in a multiplicity of interconnecting processes, now both biological and digital, to feel part of a cyber-organic ecology.

The immersive exhibition presented at the Louisiana Museum also incorporated a curated narrative of ecoLogicStudio's broader research ecosystem. This included GAN-Physarum (2022), a machine-learning slime mold trained to dream up blue-green city plans, which was first exhibited in Paris at the Centre Pompidou and is now part of its permanent collection. Also featured was Tree.One (2024), a living photosynthetic 3D-printed sculpture—which was also shown in South Korea at the Hyundai Motorstudio in Busan and Seoul, and other exhibition venues—alongside the biodesign experiments of the Synthetic Landscape Lab and the Bartlett School of Architecture's Urban Morphogenesis Lab in London.

The Louisiana Museum installation demonstrated that architecture is not merely responsive, it can be regenerative. Through the micromanagement of photosynthesis, the symbiosis of algae and fungi, and the circular reuse of carbon and biomass, Deep. Forest transformed the gallery into an ecological machine. It presented a prototype for post-anthropocentric design: a space where environmental intelligence is not simulated but embodied.

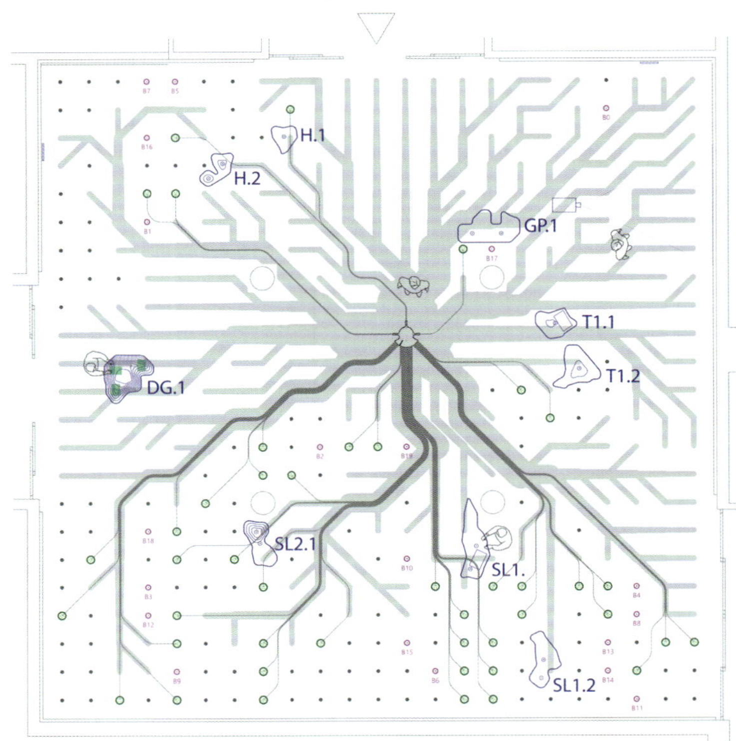

Reimagining Domesticity Through Microbial **Architectures**

This embodiment took a new turn at the 2025 Triennale Milano International Exhibition, with DeepForest[3] tackling the question of domesticity. Where its predecessor at the Louisiana Museum framed architecture as a synthetic forest, this subsequent iteration compressed the forest into an intimate domestic scale that reimagined the very idea of home as a living, microbial ecosystem.

In DeepForest[3], the forest did not surround the dwelling; it became itself a place to dwell. The floor was an engraved infrastructure, the walls were a mycelial forest, and the air was filtered by photosynthetic organisms. Blurring the boundary between interior and exterior, natural and synthetic, human and microbial, the installation proposed a new kind of domesticity that remained permeable, intelligent, and coevolving with its inhabitants.

The space was framed by three types of architectural component, each an evolved descendant of the Louisiana prototypes. Photosynthesizers, filled with cyanobacteria, actively captured CO_2 supplied within the gallery environment via a floor-integrated

piping network. These glass vessels were arranged to form a breathable membrane, a wall that became a porous filter, alive with metabolic activity. Biodegraders, built from 3D-printed bark-like shells made of algae biopolymers, hosted living mycelium and marked the terminals for the floor network, engraved in cork and bifurcating like giant mycelium hyphae (tubular filaments). Carbon storers, hosting active lichen colonies, finalized the partitioning system.

Completed with two Zola benches and an AIReactor desktop photobioreactor, DeepForest[3] was a tangible prototype for biotechnological living grounded in the metabolic cycles of algae and fungi, but one expressed through digital design and material computation. The systems that animated it, like the bubbling algae, the growing mycelium, and the fluctuating CO_2 levels, were made visible and became expressive. As such, DeepForest[3] turned sustainability and circularity into a lifestyle interface promoting a radical cultural reframing.

Sustainability ceases here to be an invisible infrastructure of accounting platforms. Instead, it is lived, sensed, and celebrated. In this context, domestic space, and domesticity, is a place of collaboration with nature, where intelligence is distributed across microbes, machines, and human minds alike, affording a vision for how our future homes may breathe, grow, and metabolize alongside us.

Toward a **Cyber-Organic** *Future*

As the climate crisis intensifies and urban populations grow, the architectural profession faces a pivotal choice: continue designing static structures of control for humans to resist ecological change or conceive new dynamic, living systems that can coevolve with the environment. The Deep.Forest project argues emphatically for the latter.

In the biodigital forest room and the domestic microbial habitat featured here, ecoLogicStudio crafts immersive environments where AI, biotechnology, and architecture coexist. These are not speculative fantasies but actionable prototypes leading to real-world projects like the Design Apothecary in Turin, Italy (2025), a studio and research space with its own air-purifying algae garden, demonstrating how buildings can breathe, sequester carbon, process waste, and regenerate resources, all while enriching spatial experience.

At the core of this work lies a design ethos where beauty emerges from symbiosis, as an expression of ecological intelligence. The adoption of living ornament, where metabolic processes are revealed and celebrated as aesthetically significant, marks a departure from the often-sterile language of greenwashing, inviting us to develop emotional and cognitive relationships with the ecosystems we coinhabit.

Furthermore, the collaboration between architectural practice and academia underscores the importance of transdisciplinary research in realizing such visions, pointing to the emergence of a new industry. The convergence of science, design, computation, and philosophy is what allows these prototypes to move beyond installations and become living embodiments of a post-anthropocentric architecture. Growing, cultivating, dreaming, endowed with artificial intelligence: this more-than-human architecture operates within a larger web of life, which includes humans alongside bacteria, fungi, algae, and machine learning algorithms. It does not offer control or certainty about the future, but does offer resilience, adaptability, and beauty born of complexity.

It is an architecture that negotiates, metabolizes, and evolves. And, perhaps most importantly, it is an architecture that learns.

← **ecoLogicStudio, DeepForest[3], *We the Bacteria: Notes Toward Biotic Architecture* exhibition, 24th Triennale Milano International Exhibition, Milan, Italy, 2025**

Visible systems turned the walls into a living, cyber-organic laboratory; the walls and floor were trees and roots, integrating microbial intelligence in a living, permeable architectural ecosystem, while air pumps circulated air and CO_2, supporting algae and mycelium growth.

↗ The DeepForest[3] interior space invited visitors to engage with and inhabit a living microbial ecosystem: sitting on a Zola bench composed of a honeycomb cardboard base and modular cork blocks that served as a breathable substrate for the cultivation of living mycelium cultures, and breathing fresh oxygen produced by the photosynthesizers and the AIReactor in action.

Genome to G-Code

Milad Showkatbakhsh and Michael Weinstock

Bio-Intelligence in Emergent Architecture

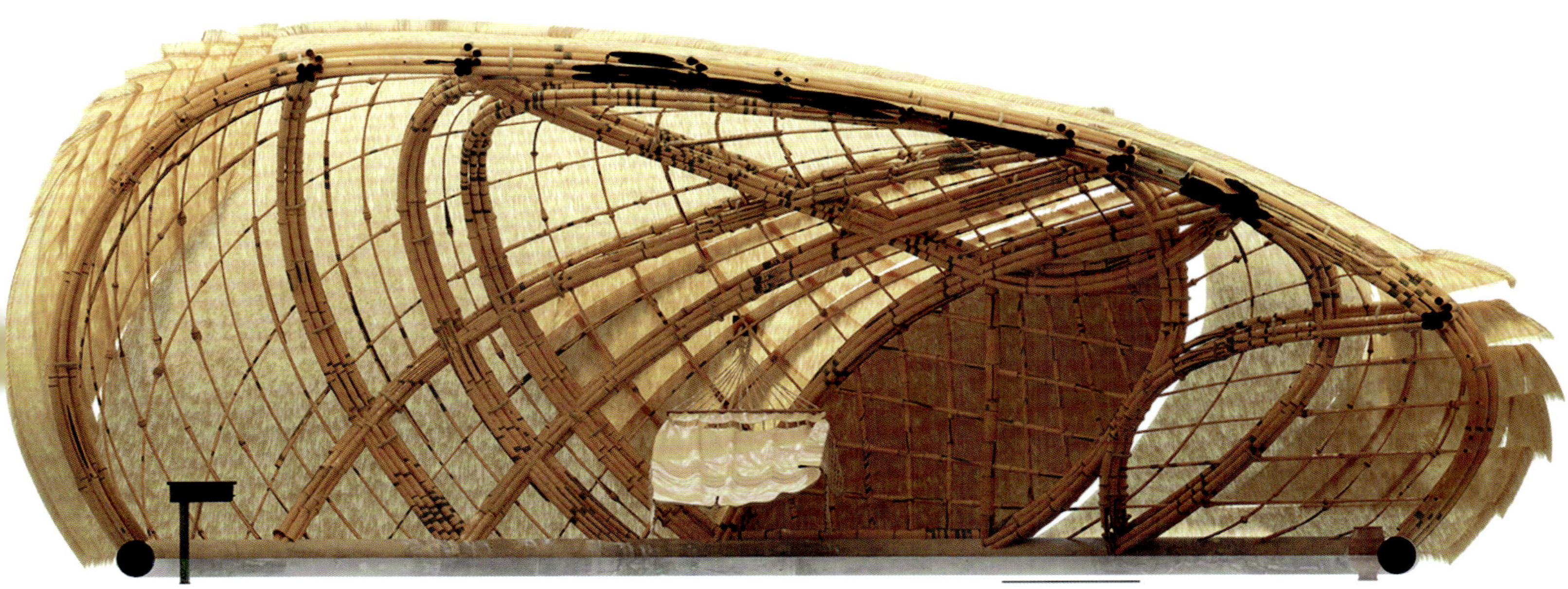

↰ **Luis Bosoms Hernandez, Rodrigo Maron Rius, and Maxence Fromentin, "Maak Bo'oy," Emergent Technologies and Design [EmTech] Master of Architecture dissertation, Architectural Association School of Architecture, London, 2021–22**

This research develops a hurricane-resilient, community-based dwelling system for the Maya people through its integration of wind-driven, form-finding, local bamboo materials and vernacular architectural principles. Beyond new housing morphologies, it proposes a micro-urban and regional strategy to create safer, economically sustainable evacuation networks across Mexico's Yucatán Peninsula.

← **Anna Maria Oldakowski, Manya Singhal, Anastasiya Katliarskaya, and Ziyue Gao, "Biobot," Emergent Technologies and Design [EmTech] Master of Architecture dissertation, Architectural Association School of Architecture, London, 2021–22**

This thesis proposes a green network to reconnect London's fragmented ecosystems by deploying adaptive "biobot" modules—self-sustaining ecological machines that regenerate underused spaces. Through biodesign and real-time environmental feedback, these biobots merge architecture, energy production, and ecological restoration, fostering new human-nonhuman relationships and climate resilience.

A persistent paradox has emerged in recent architectural discourse: Although digital technologies have expanded formal and computational capabilities, the discipline remains limited in its response to the escalating ecological crises. Biodesign—the integration of principles derived from living systems with design culture—has emerged in response to this paradox, proposing artifacts that perform as active ecological agents rather than as passive envelopes. The Emergent Technologies and Design programme, or [EmTech], at the Architectural Association School of Architecture, in London, is a pioneering laboratory dedicated to the integration of biological intelligence, computational methodologies, and advanced fabrication techniques. [EmTech] develops innovative theoretical and material paradigms, contributing to the reshaping of contemporary discourses in architecture and urban design. Treating built form, material behavior, and data flow as interconnected elements within a single cyber-physical system, [EmTech]'s foundational claim is that biological matter and information are equally important design tools, capable of generating architecture with the resilience and circularity required by today's environmental challenges.

Founded upon the thesis that nature and artifice are strongly coupled, [EmTech]'s research trajectory has progressively developed and articulated a bioinformatic paradigm wherein material systems, architectural forms, and urban fabric are re-envisioned as adaptive, mutable, and information-rich ecological systems. These systems continuously recalibrate their behavior across diverse scales and interfaces. Supported by an expanding community of interdisciplinary researchers, [EmTech] rigorously explores how biological intelligence can inspire and inform novel architectural strategies, particularly in response to contemporary global challenges. The studio seeks to integrate biological principles with advanced computational design, advocating for ecological sensitivity and adaptability within architectural practice. The aim is pragmatic rather than speculative: to develop a replicable design platform capable of producing performative material systems, scalable from individual building components to territorial settlements. Our next-generation platform is structured to offer a variable template for future iterations of ecologically sensitive architectural design across a range of scales. Building on the core frameworks guiding [EmTech]'s evolution, we outline future strategies aligned with rapid technological progress and critical ecological demands.

Dual Ontology of the Living

Central to [EmTech]'s theoretical position is the recognition that natural systems constitute a dual ontology; that physical organisms are entwined with the informational dynamics of the biological processes of genes, feedback loops, and evolution. Our research at [EmTech] abstracts both physical and process factors and recouples them within computationally driven design workflows. We use physical experimentation to explore form, structure, and behavior, and, in parallel, convert the material information into a digital format to extract the parameters and relational logics for computational development. For example, helicoidal fiber layups, hygroscopic swelling, or growth-induced prestress can be transcribed into algorithmic rules for digital generative systems, ensuring that biological performance logics are preserved. Informational transposition translates gene regulation into parametric chromosomes, developmental processes into growth grammars, and evolutionary pressures into multi-objective fitness functions. By translating each biological abstraction into both a material protocol and an algorithmic rule set, our studio advances biodesign beyond visual aesthetics toward an integrated science of adaptive form.

↘ **Fabiana Aja, Hosein Shahhoseini, and Deniz Uluköy, "Atitlán," Emergent Technologies and Design [EmTech] Master of Architecture dissertation, Architectural Association School of Architecture, London, 2023–24**

Reimagining Xochimilco's ancient *chinampa* system through modular, porous hexagonal structures made from sustainable materials, this thesis aims to restore aquatic ecosystems disrupted by urbanization. Integrating Indigenous knowledge with modern technologies like filtration systems and environmental simulations, the project proposes a resilient urban fabric that fosters ecological restoration and sustainable development in Mexico.

Evolutionary Computation as a Design Model

While conventional optimization relies on deterministic methods to reach a predefined goal, our studio employs evolutionary computation—such as genetic algorithms, agent-based models, and coevolutionary search—as a core epistemological framework. Natural selection's ability to explore complex fitness landscapes is reinterpreted as a design method capable of balancing multiple objectives simultaneously—such as structural efficiency, daylight performance, embodied carbon, sociocultural factors, and circular material systems. Within this framework, architectural species are bred rather than drawn; authorship resides in curating search spaces and fitness regimes rather than in sketching forms. The significance for biodesign is twofold. First, evolutionary algorithms formalize adaptability as a design driver from conception, mirroring biology's capacity for continual negotiation with changing environments. Second, the same computational framework proves scalar-independent, capable of steering with equal fidelity the evolution of a mycelium sandwich panel or a kilometer-scale logistics lattice. This scalar independence positions [EmTech]'s methodology as a transferable template for future biodesign research.

Living Materials and Cyber-Physical Craft

↑ **Ran An, Shengyao Zhang, Chuheng Tan, and Haipeng Zhong, "Re-Myce," Emergent Technologies and Design [EmTech] Master of Architecture dissertation, Architectural Association School of Architecture, London, 2022–23**

This project uses biodesign to develop landslide-resistant structures in Yunnan, China, by combining timber with mycelium composites. The mycelium layer enhances strength and resilience, offering a sustainable, eco-adaptive solution to erosion and impact forces.

→ **Anastasiya Katliarskaya, Maxence Fromentin, and Ziyue Gao, "Ecological Settlement Design in the Arctic," Emergent Technologies and Design [EmTech] Master of Science studio seminar project, Architectural Association School of Architecture, London, 2021–22**

This research designs an adaptable ecological settlement for Ilulissat, Greenland, by developing modular building morphologies and network systems that can relocate under extreme conditions. Using iterative design and evolutionary optimization tools, the project evolves flexible, reconfigurable structures to balance sustainability, functionality, and climate resilience.

→ **Burak Aydin, Mehmet Efe Meraki, Prakhar Patle, and Rushil Patel, "Gap in the Cloud," Emergent Technologies and Design [EmTech] Master of Architecture dissertation, Architectural Association School of Architecture, London, 2023–24**

In this thesis, data centers in London are reimagined as integrated, multifunctional urban spaces by combining data infrastructure with public engagement and food production through innovative material and spatial systems. Using shape-grammar-based design optimization, it proposes a new typology where data centers become active, adaptable elements of the urban ecosystem rather than isolated facilities.

Biodesign relies on material practices that can be empirically tested and validated. We therefore frame fabrication as cyber-physical craft: a continuous feedback loop in which simulation and making coevolve. Robotic sheet forming, spatial extrusion, and incremental weaving are coordinated in real time, with data dynamically adjusting toolpaths during fabrication—embedding process memory directly into the artifact. Three representative material systems exemplify this approach. A recent [EmTech] dissertation, "Re-Myce," explored using mycelium for restoring landslide-impacted areas. Mycelium biocomposites thickened in regions of predicted impact, producing strain-hardening barriers for debris-flow mitigation. Another project, "Maak Bo'oy," developed hurricane-resilient bamboo lattice systems for Mayan dwellings, integrating wind-driven, form-finding and hygroscopic behavior in locally sourced bamboo to lock structural stability—echoing growth-induced reinforcement in plant stems. In a third, a green network deployed adaptive "biobot" modules—self-healing ecological machines that translate biological repair into masonry adaptation, merging material intelligence with environmental feedback.

In each case, living or bio-derived matter is treated not as a symbolic alternative to conventional materials but as a programmable medium whose responsiveness continues after construction and remains contextually active. The result is a repertoire—of self-assembly, self-repair, self-actuation—that extends the field of biodesign from speculative narrative to implementable technology. The regulation of natural systems in response to environmental change is often overlooked due to its mismatch with the human lifespan—and sometimes attention span! Our studio not only identifies these regulatory feedback loops and their role in maintaining and adapting natural systems as key drivers for biodesign, but is actively developing their computational simulation.

Metabolic Urbanism **and Multiscalar Integration**

At [EmTech], we reject the idea that biodesign must remain confined to the laboratory scale. By combining agent-based dynamic behaviors, landscape metabolism, and evolutionary search, our studio formulates metabolic urbanism as a design logic that has the capacity to negotiate territorial flows of energy, nutrients, and logistics. Three recent design research projects demonstrate this logic across a range of scales and geolocations. "Ecological Settlement Design in the Arctic," for example, weaves permafrost dynamics, freight corridors, and sea-ice recession into settlement fabrics that redistribute loads as the stability of the ground shifts. The project "Gap in the Cloud" couples server heat and CO_2 exhaust with hydroponic cultivation, converting infrastructural liabilities into metabolic assets. And "Atitlán" proposes a reimagining of porous *chinampa* fabrics to reactivate the volcanic-ash islands of Mexico City, using zeolite pores that purify eutrophic waters while supporting aquatic crop cultivation. These designs are early prototypes that explore and develop workflows grounded in biological principles and evolutionary computation, scaling seamlessly from component to city, and they each contribute to [EmTech]'s biodesign agenda of architectural design for bioclimatic stewardship.

Toward an Evolutionary Biodesign Discipline

[EmTech]'s contribution to architectural discourse lies in reframing biodesign as a rigorous, multiscalar methodology anchored in the dual ontology of natural systems and operationalized through advanced computation and cyber-physical prototyping. By publishing both code and matter, we align architectural research with the norms of reproducible science, moving the field from metaphor to method. In a period defined by climatic volatility and material scarcity, our studio asserts that architecture must evolve—literally and methodologically—within the wider biosphere it inhabits. Our studio's genome-to-G-code framework offers one possible blueprint for that evolutionary future. In this paradigm, architecture is not an object imposed upon nature but a participant within it—capable of sensing, adapting, and evolving alongside the systems it inhabits.

Microalgae Building Enclosures

Kyoung Hee Kim

Designing for Regenerative Architecture

↖ **Kyoung Hee Kim, Biochromic Window (detail), University of North Carolina at Charlotte (UNC Charlotte), North Carolina, 2021**

The interlocking X-modules function as separate bioreactors, allowing for the cultivation of different algae strains when needed. They maintain a stable environment that supports a steady increase in cell density over time.

Communities face a dual challenge of decarbonizing the built environment and promoting resilient infrastructure against climate stresses. At the same time, the rapid growth of artificial intelligence (AI) workloads is driving a surge in data center electricity demand that threatens to strain power grids. Algae-derived biodiesel, a promising third-generation biofuel, has long been considered for its environmental benefits compared to food-based biodiesel. An integrative solution is an AI-enhanced algal façade—in development at the Innovation Barn, a community hub for circular economy practices in Charlotte, North Carolina—that cultivates lipid-rich microalgae on building enclosures for biodiesel production while sequestering carbon. This regenerative technology can address persistent, sustainable carbon-negative energy to help power data centers, while transforming buildings into resilient, self-sustaining infrastructure.

Advances in algae façade design are now improving biodiesel yields and the symbiotic coupling of these systems with AI-driven data centers (using waste heat and carbon dioxide to boost algal growth and energy feedstock output). Through global architectural integration, AI-enhanced algal façades could turn the built environment into a network of bioenergy generators and carbon sinks—a strategy for grid resilience and climate-adaptive communities. As such, regenerative architecture—introduced here as a broad conceptual framework—is advanced by the integration of microalgae façades, which represents prospects for implementation while also addressing the technical and economic challenges ahead. The forthcoming case study can provide a real-world test bed for these ideas.

Carbon-Neutral Resilience

Biodiesel derived from microalgae has attracted interest as a sustainable alternative to biodiesel from food sources, for example, due to algae's rapid growth and high lipid content compared to terrestrial crops.[1] Algal strains can theoretically yield an order of magnitude more oil per hectare than crop sources, and they actively sequester CO_2 during cultivation, offering carbon-negative potential if deployed at scale.[2] Ongoing challenges, however, have prevented algae biodiesel from becoming economically viable. Photobioreactor systems require significant capital investment and operating energy, mainly for aeration and harvesting, resulting in biomass production costs of $4.50–$11.80 per kg in closed bioreactors, far above the $0.40–$2.10 per kg achievable in open ponds.[3]

There are building sectors where biodiesel remains indispensable. Aviation is a prime example due to the lack of viable electrification or alternative fuels. In addition, emergency backup power, farm equipment, and off-grid infrastructure are other critical domains. Diesel generators remain the ubiquitous choice for backup power in hospitals, data centers, and remote communities due to their reliability and energy-dense fuel. These sectors' reliance on diesel underscores the importance of improving biodiesel's cost and production scalability.

At the same time, the explosive growth of digital infrastructure is creating new energy demands. The International Energy Agency projects that worldwide data center electricity demand will more than double by 2030, exceeding 940 TWh, driven chiefly by AI workloads.[4] This soaring demand risks overstressing electrical grids, and many regions are already grappling with how to provide reliable power for new data center clusters and communities, especially during extreme weather events. The need for resilient power to ensure a 100 percent operational time means that most data centers maintain banks of diesel backup generators, further reinforcing the use of diesel fuel for critical infrastructure.

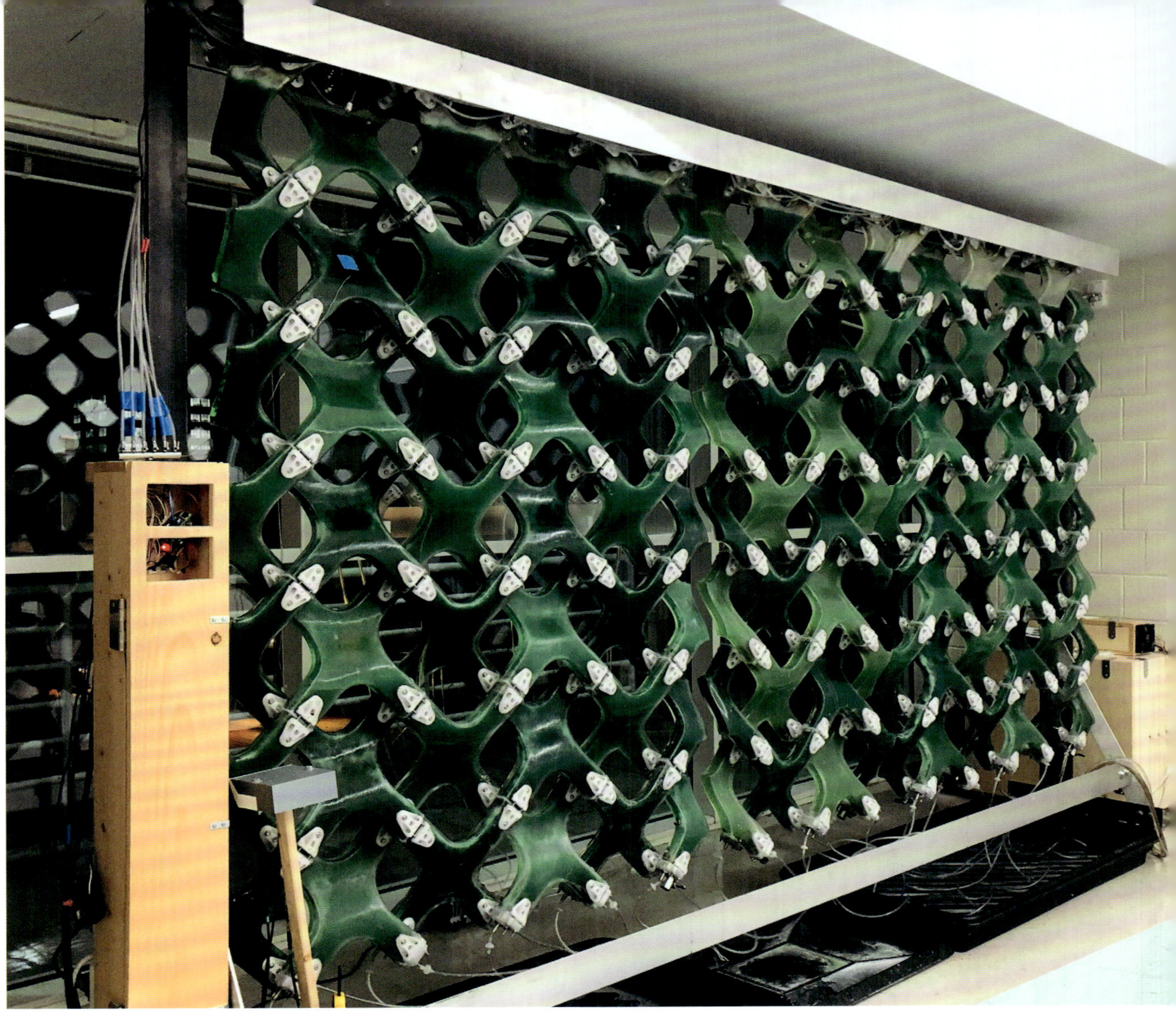

→ **Kyoung Hee Kim, Biochromic Window, UNC Charlotte, Charlotte, North Carolina, 2021**

The biochromic façade incorporates X-shaped modules that allow a view out for occupants and operate as photobioreactors for microalgae cultivation, fostering a symbiotic relationship between human and nature.

↓ **Kyoung Hee Kim, integrated design process of Biochromic Window, UNC Charlotte, Charlotte, North Carolina, 2021**

The integrated design process highlights how performance analysis, material properties, and fabrication methods inform the overall system design. The modular construction strategy enables flexible scalability across various façade sizes and geometries, supporting both functional adaptation and practicality.

Biochromic Microalgae Façades

A full-scale prototype of the biochromic microalgae window installed at the University of North Carolina at Charlotte (UNC Charlotte) in 2021 intakes carbon dioxide along with sunlight and generates oxygen-rich air, which is released from the top of the system. The system reduces building cooling loads by dynamically changing the microalgae cell density depending on sunlight intensity and by regulating incoming solar radiation. For example, a microalgae system filled with a dense algae culture blocks solar penetration more effectively than one filled only with clear water, indicating that such "living" windows can dynamically modulate solar input based on algae growth. This behavior directly supports the energy-saving principle of the system. High algae density in summer conditions can cut down cooling loads by shading excess sunlight and absorbing heat, whereas a lower density would allow more sunlight through in winter seasons. Simultaneously, this improves the indoor environmental quality by reducing glare and enriching indoor air with oxygen through photosynthesis, thereby enhancing occupant wellness.[5]

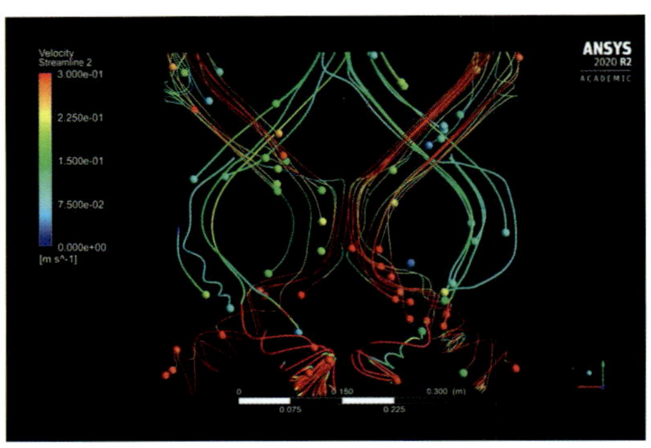

The Biochromic Window was implemented through an integrated design process that bridges architectural design and biotechnological innovation. The system begins with multifunctional modular design using performance analysis to optimize structural integrity and environmental responsiveness. Fabrication was realized through rotocasting techniques, producing durable, transparent modules shaped for both aesthetics and functionality. These modules were assembled into interlocking microalgae façade systems, which support microalgae cultivation in natural indoor environments. The system serves multiple functions, including solar shading, air purification, and carbon sequestration, making the façade not only functional and aesthetic but also regenerative. The interlocking X-shaped modules provide a view outside, for better user experience, and offer scalable flexibility to accommodate a wide range of window shapes and configurations.

← **Kyoung Hee Kim, Crown Shyness Window, Innovation Barn, Charlotte, North Carolina, 2024**

This AI-driven microalgae façade, inspired by the crown shyness phenomenon, dynamically adjusts cultivation conditions, optimizing carbon capture and biodiesel feedstock production, while building cooling load reduction.

→ The Crown Shyness Window features a continuous channel—drawing inspiration from the natural crown shyness phenomenon—and distinct crown zones composed of concave and convex geometries. The continuous crown shyness channel facilitates the movement of liquid and air across these photobioreactor crowns, enabling effective cultivation.

AI-Enhanced Microalgae Façades

The confluence of slow progress in biodiesel scalability and skyrocketing hyperscale AI energy use calls for innovative solutions that address both challenges together. As a sustainable alternative, AI-enhanced algae façades have emerged as a compelling nature-based solution at this nexus, with the goal of turning data centers themselves into microalgal biofuel reactors that capture carbon and produce renewable energy.

Both biomass growth and lipid content are influenced by a complex interplay of factors that determine cultivation environments including daylight intensity, nutrient level, medium temperature, CO_2, and pH. Typically, high light combined with abundant nutrients promotes high biomass growth, whereas high light paired with strategic nutrient manipulation improves lipid yield. Light intensity moderately enhances biomass but only weakly affects lipid levels at high intensities.[6] CO_2 enrichment (above ambient 0.04%) positively correlates with biomass production and slightly promotes lipid accumulation, especially when combined with nutrient stress.[7] The effects of temperature and pH are milder, consistent with general algal stress-response patterns. Responsive to these differentials, AI-monitored microalgae façades automatically adjust cultivation factors to maximize growth and environmental benefits.

The Crown Shyness prototype (2024) installed at the Innovation Barn, in Charlotte, North Carolina, was inspired by the natural phenomenon of crown shyness, where reciprocal symbiotic relationships create voids that allow sunlight to reach lower vegetation while protecting individual crowns from the widespread transmission of diseases. The Crown Shyness Window features a continuous crown shyness channel, and distinct crown zones composed of concave and convex geometries. The concave crowns at eye level offer an unobstructed view out, while the convex crowns at the top and bottom function as photobioreactors for growing microalgae. The continuous crown shyness channel enables the movement of liquid and air across the panel, ensuring effective and evenly distributed cultivating environments. The complex geometry of the panel was achieved through a thermal forming process in which polycarbonate sheets were heated and slumped onto a custom-machined aluminum mold. This fabrication method allowed for precise replication of the dynamic form, optimizing both aesthetics and functionalities. The AI-enabled system continuously monitors and optimizes the complex interplay of environmental parameters in real time, reinforcing the façade's role as a living, regenerative building skin.

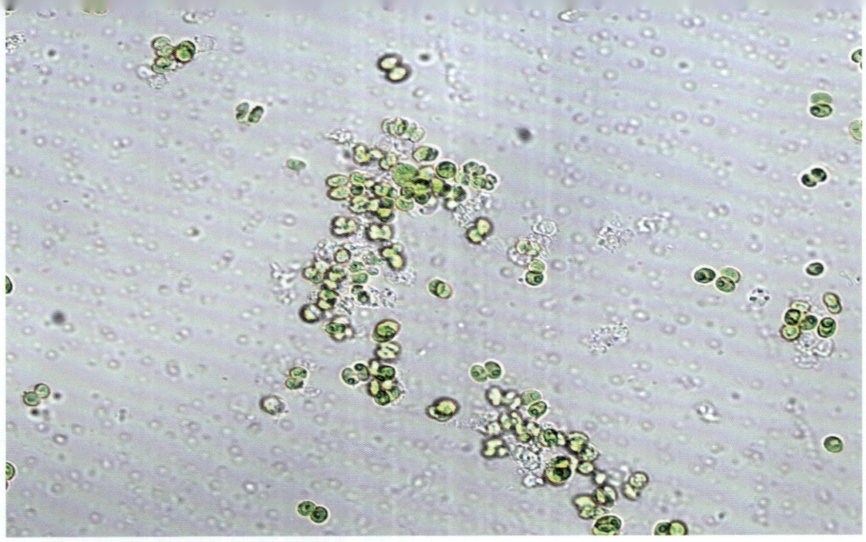

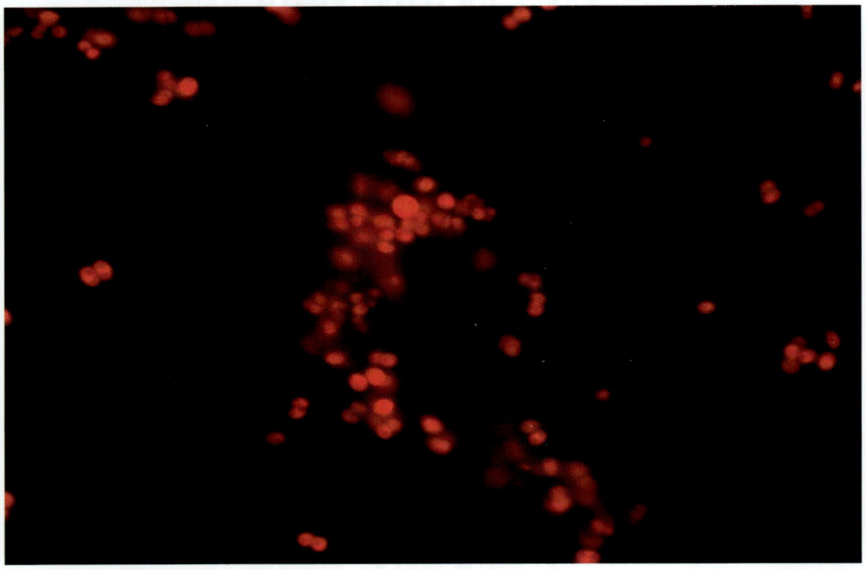

← **Kyoung Hee Kim, lipid fluorescence analysis of *Nannochloropsis*, UNC Charlotte, Charlotte, North Carolina, 2025**

Nile red fluorescence microscopy shows bright, localized lipid bodies in *Nannochloropsis* cells. The Nile red procedure is nondestructive and time efficient.

→ **Kyoung Hee Kim, lipid fluorescence analysis of *Chlorella*, UNC Charlotte, Charlotte, North Carolina, 2025**

Chlorella vulgaris displays a more diffuse, weaker lipid fluorescence, indicating a relatively lower lipid yield under similar conditions. The procedure remains nondestructive and time efficient for comparative screening. Fluorescence imaging and microscopy support was provided by Constance Sartor and Matt Parrow at UNC Charlotte.

Nature-Based Solution

A symbiotic, closed loop is established whereby the data center's waste heat and CO_2 are directed into the bioreactor to fuel thermophilic algae growth, and the resulting algae-derived biofuel, in turn, powers the data center's backup systems or the local grid, closing a carbon loop. Additionally, the waste heat can be used for drying harvested algae sludge, saving energy for the dewatering process. It is also important to understand how traditional freshwater algae such as *Chlorella vulgaris* and *Nannochloropsis* respond to biomass and lipid yield under different cultivation environments. *Chlorella vulgaris* is a freshwater green spherical-shaped microalga, and a good candidate for biofuel feedstock production due to its ability to accumulate high lipids when stressed.[8] In addition, *Nannochloropsis*, a brackish and marine species, possesses a remarkable lipid accumulation ability under nutrient restriction.[9]

Biological performance metrics can be determined by microalgae growth rate and productivity. Microalgae growth rate refers to the speed of cell multiplication (g/L/day), while productivity indicates the total lipid output (%/g biomass). It is possible to have a high growth rate but low productivity if the cells are small. Likewise, a slower growth rate can result in high lipid productivity. Lipid productivity potentials can be measured by Nile red techniques, a fluorescent dye that can be used to visualize neutral lipids, such as triacylglycerides, in microalgal cells. The Nile red procedure is widely used because of its nondestructive nature, accurate results, and time efficiency.[10] Microscopy images of microalgae were produced at UNC Charlotte to study their lipid productivity. While *Nannochloropsis* was observed to have bright red fluorescence, indicating parts of cells that contain high amounts of neutral lipids, *Chlorella vulgaris* demonstrated weaker but more spread-out fluorescence signals, rather than localized signals, suggesting a relatively lower lipid yield.

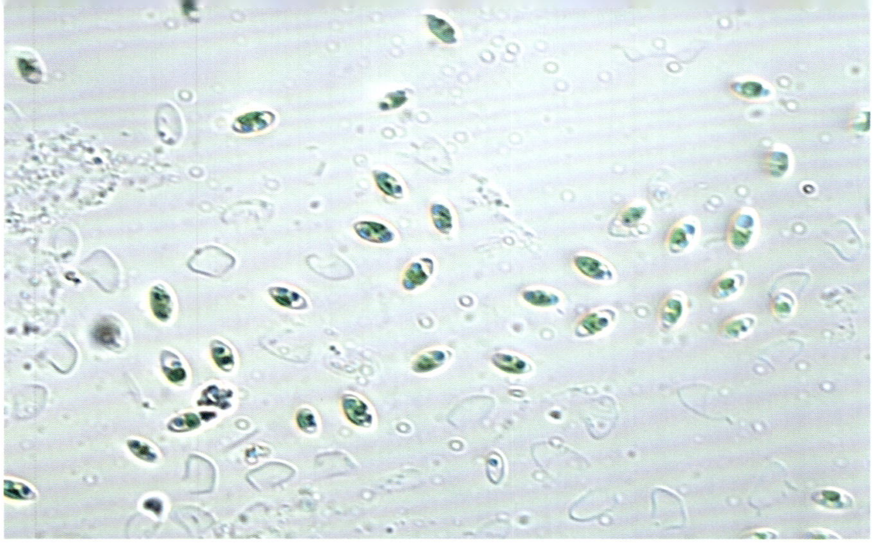

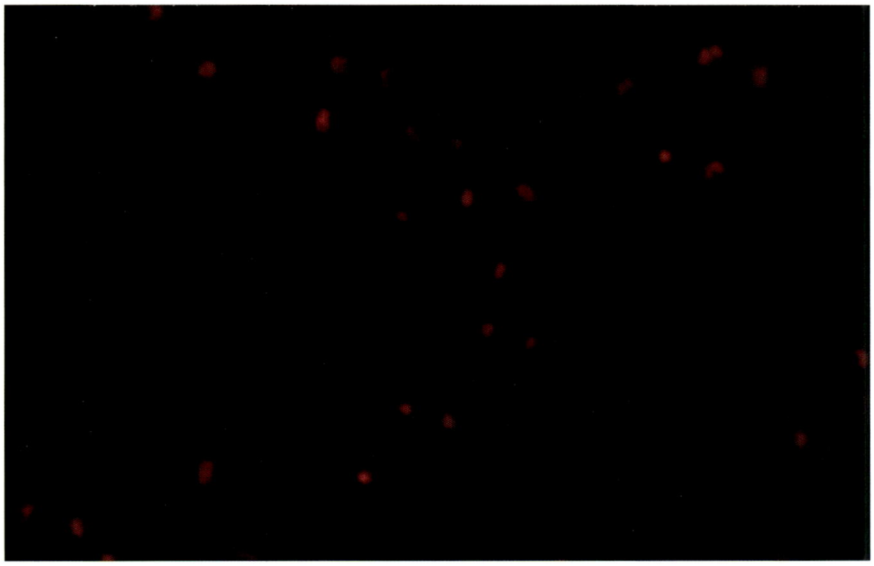

1 Livia M. Casanova, Leonardo Mendes, Thamiris Corrêa, et al., "Development of Microalgae Biodiesel: Current Status and Perspectives," *Microorganisms* 11, no. 1 (2023): 34.
2 Mladen Bošnjaković and Nazaruddin Sinaga, "The Perspective of Large-Scale Production of Algae Biodiesel," *Applied Sciences* 10, no. 22 (2020).
3 Casanova et al., "Microalgae Biodiesel," 34.
4 "Electricity 2024: Analysis and Forecast to 2026," International Energy Agency (Paris), January 2024, https://www.iea.org/reports/electricity-2024.
5 Kyoung Hee Kim, *Microalgae Building Enclosures: Design and Engineering Principles* (London: Routledge, 2022).
6 Maryam Al-Qasmi, Nitin Raut, Sahar Talebi, Sara Al-Rajhi, and Tahir Al-Barwani, "A Review of Effect of Light on Microalgae Growth," in *Proceedings of the World Congress on Engineering*, vol. 1 (London: International Association of Engineers, 2012), 608–10.
7 Caroline Montoya-Vallejo, "Biomass and Lipid Production by the Native Green Microalgae Chlorella sorokiniana in Response to Nutrients, Light Intensity, and Carbon Dioxide: Experimental and Modeling Approach," *Frontiers in Bioengineering and Biotechnology* 11 (May 2023).
8 Gregory J. O. Martin, David R. A. Hill, Ian L. D. Olmstead, et al., "Lipid Profile Remodeling in Response to Nitrogen Deprivation in the Microalgae *Chlorella* sp. (Trebouxiophyceae) and *Nannochloropsis* sp. (Eustigmatophyceae)," *PLOS One* 9, no. 8 (2014).
9 Martin et al., "Lipid Profile Remodeling."
10 Gibrán S. Alemán-Nava, Sara P. Cuellar-Bermudez, María Cuaresma, et al., "How to Use Nile Red, a Selective Fluorescent Stain for Microalgal Neutral Lipids," *Journal of Microbiological Methods* 128 (September 2016): 74–79.

Toward Resilient, Regenerative, and Carbon-Neutral Futures

AI-enhanced algal façades represent a bold reimagining of the relationship between buildings, energy, and the environment—one that holds promise for addressing some of the most pressing challenges of sustainable, carbon-neutral development. By leveraging living microalgae systems on architectural skins, powered and optimized by AI integration, tomorrow's buildings could become active contributors to energy resilience and carbon sequestration. Moreover, symbiotic integration with AI data centers could tackle the growing electricity strain of our digital economy, especially during extreme weather events. An AI-algae façade on a data center would feed on the center's waste heat and CO_2—for energy savings and carbon sequestration—while also producing sustainable biodiesel feedstock, effectively creating a localized, resilient, circular energy ecosystem.

Significant challenges remain in achieving scalability and commercialization readiness. A holistic life-cycle assessment is necessary to fully understand the total resource and energy inputs, along with resulting environmental impacts associated with system fabrication, assembly, transportation, operation, maintenance, and end-of-life management. The system longevity of algal façade systems over building lifespans must be proven, as reliability is paramount if they are to power critical facilities. Economic viability will depend on continued declines in sensor, automation, and cultivation costs—progress that is likely as AI and biotech mature. Both automation and operational efficiency are important to scale up production to meaningful fuel volumes in urban environments. It is also essential to have supportive policy and further research and development, including pilot projects in real-world settings to provide visible proof of concept to stakeholders.

The convergence of AI, renewable biofuels, and regenerative architecture points toward carbon-negative, grid-resilient communities, where every building is a source of power and every façade a canvas for carbon capture. By turning our cities into photosynthetic power plants, we can not only mitigate climate change by curtailing CO_2 but also create urban infrastructures that gradually adapt to evolving climate stresses and societal needs. If successful, AI-enhanced algal architecture could become a cornerstone of sustainable cities—living proof that our built environment can evolve from a passive backdrop into an active agent of ecological regeneration.

DESIGNING WITH POLYCULTURES— A CONTINUUM OF BIORECEPTIVITY, POIKILOHYDRY, AND BIOREMEDIATION

Marcos Cruz and Brenda Parker

Embedding biological systems in our built environment offers the potential to mitigate our anthropogenic footprint while reducing levels of pollution. Bio-integrated design is a regenerative practice whereby species and/or bioprocesses form intrinsic parts of building designs, while their interaction with wider ecological networks is inherently considered. Integration implies giving biology its full agency without necessarily coercing it to exclusively serve our human needs, generating hybrid conditions that are "partly designed objects and partly designed living material."[1] Fundamental to this is a time-based materiality that is receptive to growth, creating microrefugia where life can proliferate due to the availability of water and favorable biotic conditions. The plasticity and robustness exhibited by assemblages of organisms provides vital ecological connectivity between humans and architecture.

Early Bioreceptive Explorations

↰ **Shneel Malik, Brenda Parker, and Marcos Cruz, INDUS 1.0, Brompton Design District, London Festival of Architecture, London, 2019**

Demonstrator wall with algae-laden hydrogel extruded on handcrafted tiles for bioremediation of water.

↗ **Marcos Cruz in collaboration with Javier Ruiz and Richard Beckett with manufacturing by Pennine Stone, Poikilohydric Living Wall, London, 2019**

Bioreceptive wall detail with transplanted moss in preparation at UCL at Here East, London, for exhibition at *La Fabrique du Vivant*, Centre Pompidou, Paris, 2018.

→ **Marcos Cruz in collaboration with Haddonstone, Poikilohydric Living Wall, Merchiston Park, Edinburgh, 2023**

Inchoate phase of biocolonization on a northwest-facing façade wall.

→ **Marcos Cruz, Poikilohydric Living Wall, Merchiston Park, Edinburgh, 2023**

Evidence of algae growth on bioreceptive material after first winter cycle.

Introduced by the Belgian agricultural engineer Olivier Guillitte in 1995, the concept of bioreceptivity was first used in the context of building conservation to describe "the ability of a material to be colonized by living organisms."[2] It was later expanded by researchers in conservation science to include the colonization of built cultural heritage within submerged and subsoil environments.[3] There are several site-specific realms of bioreceptive activity that present opportunities for design interventions: aquatic (e.g., coralline algae propagation in marine environments); subterranean and subsoil (i.e., building foundations and dynamic terrestrial conditions in soil crusts); and epilithic surface or subaerial bioreceptivity (i.e., biofilm formation on stone or cementitious contexts), the latter underpinning the projects featured here.

Designed by the experimental practice marcosandmarjan, the Algae-Cellunoi wall[4] was an installation at the 2013 *ArchiLab* exhibition in Orléans, France, which comprised a three-dimensional wall with integrated laboratory-cultured dishes intended to host outdoor growth. Soon after, in 2014, the Spanish biologist Sandra Manso concluded a doctoral research project in which she developed subaerial bioreceptivity for building applications focused on calibrating pore size, surface roughness, and pH levels on small-scale cementitious material samples to propagate moss.[5] These precedents informed the later definition of "bioreceptive design," understood as the intersection of biotic and abiotic parameters to enable a systematic biocolonization of buildings and urban infrastructures.[6] Material and morphological variation and the intersection of environmental parameters, including moisture, light, nutrient level, and surrounding vegetation, were used to enhance the establishment of synthetic biotopes on material substrates defined by physical and biochemical properties. Since 2016, these principles have been extended by Marcos Cruz in the design of a series of Poikilohydric Living Walls.[7]

Poikilohydric Living Walls

Poikilohydry refers to the aptitude of certain plant species for survival under fluctuating conditions of moisture. The unique physiological capacity of most cryptogams—algae, mosses, and lichens—enables them to resist lengthy drought periods and extreme temperatures. Able to switch on and off their photosynthetic activity, these species rehydrate by absorbing rainwater, vapor, and dew. Poikilohydric Living Walls were thus conceived to improve the environmental and aesthetic quality of cities without relying on mechanical irrigation and human maintenance. An underlying concept here is that of architectural "bark," which, in contrast to an architectural protective "skin," implies the hosting of an ever-changing ecology of species.[8] Poikilohydric Living Walls were created as ornate surfaces that have the capacity to augment greening by capitalizing on the natural weathering process of buildings. Extensive research was developed in three different phases and places.

The first period (2016–18) focused on an outdoor environmental trial at the Bartlett School of Architecture, University College London (UCL), in which magnesium phosphate concrete (MPC) and ordinary Portland concrete panels were compared, while water runoff and thermal variability was tested through a full annual cycle. The experiments proved that morphological variability is key to retain water and accelerate the establishment of cryptogamic growth. A further period focused on the design of new surface patterns, as well as on the development of manufacturing techniques and transportation of glass-reinforced concrete panels, Porous TecCast, for the exhibition *La Fabrique du Vivant* at the Centre Pompidou in Paris (2019) and a freestanding wall in a primary school in Vauxhall, South London (2021).

In a later stage, a series of panels were manufactured with a dry concrete mix for a house extension at Merchiston Park in Edinburgh, Scotland, where they were fixed for the first time as façade cladding in a building (2022).

Carbon and Nitrogen Uptake

Subaerial bioreceptive designs may have multiple aims; for instance, the uptake of carbon dioxide, nitrogen dioxide (NO_2), and particulate matter for air quality improvement. Even though estimations suggest that cryptogamic uptake of CO_2 corresponds to 7 percent of overall net primary production by terrestrial vegetation, and up to 50 percent of nitrogen fixation on land,[9] it is challenging to extrapolate global ecosystem estimations and apply these to local concepts. The contribution of carbon uptake on bioreceptive surfaces is difficult to measure due to the scattered occurrence of cryptogams on external walls. Their growth is defined by daily and seasonal cycles that create growth variations in which the poikilohydric properties of these miniature ecologies absorb and release similar amounts of carbon.[10] Contrary to speculation in earlier research phases, and as often proclaimed in other projects of this nature, the quantifiable benefits of carbon uptake of moss on bioreceptive walls are far fewer than initial projections. On the other hand, acoustic attenuation and/or thermal control examinations show encouraging results but require further testing.

Bioremediation by Assembled Communities

The tremendous surface area that humans have created through the construction of buildings and infrastructures holds promise for the environmental impact of bioreceptive façades in terms of storm water management, if consistently applied in future cities. Yet there is a further dimension when considering the legacies of our Anthropocene: the removal of pollutants from water. Biological processes offer the potential for uptake of heavy metals or degradation of organic molecules that contaminate surface runoff. An outdoor demonstrator, INDUS 1.0, installed during the London Design Festival in 2019, expanded upon investigations of extrusions of microalgae immobilized in hydrogel to create an activated medium for bioremediation.[11] Ceramic tiles were glazed to direct water onto the algae-laden matrix in order to sequester pollutants and allow for their removal.

The bioreceptive activity of porous ceramics, however, remained little understood. Subsequent experimentation also suggested that mixtures of algae and bacteria create more stable and amenable consortia for environmental performance,[12] which led to the idea that bioreceptive scaffold design could offer additional environmental benefits.

→ **Studio Biocene, Augmented Polycultures, Museum für Gestaltung, Zürich, 2025**

Confervant phase showing intense growth of biofilm after seven months, demonstrating increased intensity where surface texture and morphology influence the colonization of microorganisms.

Augmented Polycultures

Addressing these challenges and following up previous research prompted a re-examination of how we remediate runoff and greywater in cities. Augmented Polycultures (2024), by design collective Studio Biocene, was a demonstrator on which to grow algal communities, connecting material properties for bioreceptivity with bioremediation. Conceptually, the design resembled a vertical gargoyle with embedded grotto features. The main body of the installation was made of a nonplanar robotic extrusion of porous ceramics computationally designed to increase the residence time of water and nutrients in specific areas. The control of porosity, morphology, and orientation triggered biotic and abiotic factors that govern growth at both microscopic and macroscopic levels by encouraging biofilms—emergent assemblages of algae and bacteria bound within a self-produced matrix of extracellular polymeric substances.

→ **Studio Biocene: Brenda Parker and Marcos Cruz with Nina Jotanović, Jingyuan Meng, Guillem Perutxet Olesti, and Pradeep Devadass, Augmented Polycultures, Museum für Kunst und Gewerbe, Hamburg, 2024**

Burgeoning phase, where establishment of pigmented biofilms containing algae and cyanobacteria cause the formation of a heterogenous patina after the first three months.

→ **Studio Biocene, Augmented Polycultures, Museum für Kunst und Gewerbe, Hamburg, 2024**

Gradient of glazed and unglazed layers of robotically extruded ceramic with fine sand particles, creating variable intensities of bioreceptivity. Water is supplied at variable intervals to the indoor installation via a recirculating pump housed in the slip-cast cistern as a proxy for runoff or greywater treatment.

→ **Studio Biocene, Augmented Polycultures (detail), Museum für Gestaltung, Zürich, 2025**

Microscopic imagery of the rehydrated algal consortium colonizing the porous ceramic structure after a year.

Aesthetically, Augmented Polycultures seeks to explore the possibilities of relaxing control (of purity) while retaining control (of performativity) by examining our reaction to the slimy and viscous appearance of algal scums on architectural components. By selectively defining the readability of biofilm formation, the limits of acceptance of potentially abject conditions in heterogenous biomes are exposed.

Biological Evolution and Time-Based Design

These projects have shown that the inherent bioreceptivity of materials will make architecture change over time. Against the idea of everlasting newness, weathering and biocolonization are compelling architects to design for ageing buildings. A series of biological phases have been identified through the lifespan of a design, which merit outlining here.[13] Equivalent to Guillitte's primary bioreceptivity, the *inchoate phase* describes the period when the material is new and early biofilms are being established. The rate of progression is determined by time and by prevailing environmental parameters affecting moisture availability. This eventually transitions into the *burgeoning phase*. Akin to Guillitte's idea of secondary and tertiary bioreceptivity, this second phase focuses on the proliferation of growth in regions where surface roughness or water activity is favorable. The application of additional coatings or seeding with biological material—as performed at the wall in Vauxhall—is comparable to the idea of "quaternary bioreceptivity" as defined by Patricia Sanmartín and others.[14] In a time-based design sequence, the *confervant phase* explores a moment of great intensity when the material has an abundance of well-established growth. Here, the performative aspects of the system—bioremediation, for example—are operationalized. Finally, the *biocoenosis phase* is the stage where interacting organisms live together forming a highly complex ecological community, with interspecies interactions shaped by evolutionary forces that include the decay and regeneration of specific plants. This last phase can only happen in places with higher levels of moisture availability and after a long period of time. As the most challenging to plan for, designers must relinquish control to the succession of organisms, with biodiversity assuming a multiplicity of forms that transcends the microbial or cryptogamic.

1 Marcos Cruz and Steve Pike, eds., "Neoplasmatic Design," *Architectural Design* 78, no. 6 (2008): 6.

2 Olivier Guillitte, "Bioreceptivity: A New Concept for Building Ecology Studies," *Science of the Total Environment* 167, nos. 1–3 (1995): 215–20.

3 Patricia Sanmartín, Ana Zélia Miller, Beatriz Prieto, and H. A. Viles, "Revisiting and Reanalysing the Concept of Bioreceptivity 25 Years on," *Science of the Total Environment* 770 (May 2021).

4 Algae-Cellunoi was installed at the 2013 ArchiLab exhibition, *Naturalizing Architecture*, at the FRAC Centre in Orléans, France. For more see mam, Grymsdyke Farm, and REX|LAB, ProtoRobotic FOAMing, Bartlett Design Research Folios, Bartlett School of Architecture, UCL, 2013.

5 Sandra Manso, Gemma Mestres, Maria-Pau Ginebra, Nele de Belie, Ignacio Segura Pérez, and Antonio Aguado, "Development of a Low pH Cementitious Material to Enlarge Bioreceptivity," *Construction and Build Materials* 54 (March 2014): 485–95.

6 Marcos Cruz and Richard Beckett, "Bioreceptive Design: A Novel Approach to Biodigital Materiality," *Architectural Research Quarterly* 20, no. 1 (2016): 51–64.

7 Marcos Cruz, Poikilohydric Living Walls, Bartlett Design Research Folios, Bartlett School of Architecture, UCL, 2021.

8 Cruz and Beckett, "Bioreceptive Design."

9 Wolfgang Elbert, Bettina Weber, Susannah Burrows, et al., "Contribution of Cryptogamic Covers to the Global Cycles of Carbon and Nitrogen," *Nature Geoscience* 5, no. 7 (2012): 459–62.

10 Alexandra Adelina Lăcătuşu, Marcos Cruz, Brenda Parker, and Anete Krista Salmane, "Biocentric Design: Mapping Optimal Environmental Variables for Moss Propagation on Urban Bioreceptive Surfaces," *XXVI International Conference of the Iberoamerican Society of Digital Graphics* 11, no. 2 (2023).

11 Shneel Malik, Julie Hagopian, Sanika Mohite, et al., "Robotic Extrusion of Algae-Laden Hydrogels for Large-Scale Applications," *Global Challenges* 4, no. 1 (2019).

12 Brenda Parker, Deepak Rawat, Shneel Malik, Anaëlle Vilatte, and Marcos Cruz, "A Design Manifesto for Bioremediation: 10 Principles for the Creation of Sustainable Systems for Environmental Benefit," in *AR#0 Biorémédiation* (Arles: Atelier LUMA, 2022), 13–8.

13 Brenda Parker and Marcos Cruz, "Augmented Polycultures: Scaling up Algal Ecosystems and Design of a Biofouling Aesthetic," in *Being Algae: Transformations in Water, Plants*, ed. Yogi Hale Hendlin, Johanna Weggelaar, Natalia Derossi, and Sergio Mugnai (Leiden, Netherlands: Brill, 2024), 234–67.

14 Patricia Sanmartín et al., "Concept of Bioreceptivity."

Building Application

In our cities, gradients between built and natural environments are formed by activating a roof-façade-ground biofilm continuum that exists everywhere. While construction and the "cleaning" of façades create an ongoing disturbance, bioreceptive materials can accelerate biofilm formation and thus restore the continuum on both retrofitted and new structures. By fostering poikilohydric biomes, porous materials dissolve the hard boundary of buildings, promoting a microwilderness that creates infinitesimal points of connection with the wider ecosystem. Implementing this type of greening, however, requires us to accept a biological presence on our walls in a very close proximity to us. We know that bio-integrated designed habitats can help us reduce the impact of our anthropogenic activity by sequestering pollution from water runoffs or through integration with greywater loops on vertical surfaces. While the performativity of bioremediation acts in an imperceptible manner, the resulting and fully established polycultures can offer a haptic experience that invites us to engage with a different materiality in our built environment. Polycultures produce viscous conditions that, so far, have only been appreciated in nature or in specifically designed spots like grottos. For us to promote them all around, and for us to live with and alongside such growth, requires us to shift our understanding of a new order; one that accommodates unpredictability, imperfection, and change.

Printing Biological Matter

Nancy Diniz and Frank Melendez

Cultivating Growth in Design and Fabrication

↰ **bioMATTERS, Hyperaccumulating Architectures, 2024**

Individual prototype of one of the 3D-printed clay and mycelium objects integrating locally sourced bioreceptive and biocomposite materials.

↑ **bioMATTERS, Hyperaccumulating Architectures: Micro Infrastructures for Mycoremediation Strategies, 2024**

This project was commissioned by the Museum of Architecture and Design (MAO) and Center Rog, in Ljubljana, Slovenia, and exhibited at *Future Legacies*, MAO and Center Rog, May 15–September 14, 2025. It explores the use of mycoremediation approaches to clean up heavy metal soil contamination in regions of Slovenia that were polluted due to recent, unprecedented flooding, linked to climate change.

The following projects by bioMATTERS contribute to emerging theoretical discourses on material fabrication in co-authorship with living systems. This shift embodies a broader post-anthropocentric ethos, recognizing the agency of nonhuman "actors"[1] and the capacity of computation and digital technologies to mediate and enhance complex biological interactions. The result is a more holistic and systemic approach to design, one that embraces "joint kinships"[2] between humans, machines, and biological systems, and acknowledges the interconnectedness of all life-forms and the environments they inhabit. The work foregrounds the development of hybrid fabrication ecologies—where digital manufacturing converges with biological growth. Microorganism-specific behaviors are treated as active design parameters, introducing material agency into computational workflows and prompting new inquiries into multispecies collaboration and posthuman material aesthetics.

↑ **bioMATTERS, MYCO-VASES, 2021–23**

Exhibited at *Material Matters*, the Bargehouse, OXO Tower Wharf, London, September 20–23, 2023, as part of London Design Festival 2023, these 3D-printed mycelium vases use various substrates, mycelium strains, and growth techniques.

By detailing these projects, bioMATTERS offers a multidisciplinary approach where digital tools and biological systems coalesce, creating a new hybrid space for design, and reshaping how we conceive of making itself. We employ 3D printing and robotic biofabrication not just as production tools but as coevolutionary systems that align material agency with design intent. These workflows enable the mass customization of ecologically embedded architectures, scaling the potential of bio-based materials from lab to architectural scale—without relying on extractive, high-waste manufacturing paradigms. Through this synergy, the work aims to propose not only new material aesthetics but new material economies for product design and architecture. This research is a way of merging the precision, versatility, and scalability of digital fabrication with the organic, unruly, evolving qualities of living materials. In this approach, the focus of bioMATTERS is on the interactions between digital design algorithms, biomaterials, and manual craftsmanship, resulting in the production of living materials that carry the essence of traditional artisanal practices while also pushing the boundaries of science and regenerative design practices.

Bioreceptivity and Hybridity

↑ **bioMATTERS, MYCO-ALGA, 2022–23**

This computationally designed, non-repetitive architectural tiling system, fabricated using 3D-printed mycelium-based composites and algae-based pigments, was exhibited at *Grand Designs Live* 2024, Excel, London, May 4–12, 2024, and featured as "Kevin McCloud's Green Heroes," celebrating the best conceptual and eco-friendly construction ideas.

↑ **bioMATTERS, Bacterial Patterns, 2022**

Research at bioMATTERS includes testing bio-derived alternatives to synthetic pigments and the aesthetics of 3D-printed bacteria-based pigments. This photograph illustrates the computationally designed, 3D-printed pattern, and the bacteria's emergent growth pattern.

Over the past decade, the work of bioMATTERS has evolved innovative materials approaches rooted in the concept of "bioreceptivity," which defines a material's inherent capacity to be colonized by living organisms.[3] In this context, bioreceptive materials are intentionally designed to host and sustain microbial life, allowing microorganisms to interact with, and transform, the built environment. Rather than treating a biological system as a passive resource to be exploited, we engage it as an active collaborator. For example, in our projects involving mycelium, microalgae, and bacteria, we explore how these microorganisms can co-author material aesthetics.

Through interdisciplinary workflows, our research brings together laboratory protocols for cultivating and manipulating living cells with advanced computational design strategies. It incorporates environmental simulation tools, generative modeling techniques, and robotic 3D-printing technologies, alongside expertise in microbiology and materials science. Ecological aesthetics moves away from merely biomimicking living characteristics, and instead embraces form generation as a processual, entangled, and time-based phenomenon. It foregrounds the performative, temporal, and situated qualities of living matter. In this view, aesthetics is not only visual but also systemic—manifesting through metabolic activity, growth patterns, decay, and symbiotic interrelations.

The research projects at bioMATTERS have focused on creating 3D-printable bioreceptive materials, primarily through the development of mycelium-based composites, and hybridities of mycelium coupled with other microalgae and bacteria, investigating these as pigment-producing microorganisms. For example, the MYCO-ALGA project (2022–23) integrates 3D-printed mycelium-based composites with algae-derived pigments to develop an architectural tiling system for interior design.[4] Drawing inspiration from mycorrhizal networks, variations of computationally generated branching algorithms and additive manufacturing techniques ensure each tile exhibits a unique, non-repetitive pattern and bespoke form. When assembled, the tiles generate continuously evolving surface compositions.

The research also advances methods for isolating and extracting natural pigments from fungi, bacteria, and microalgae, which are successfully applied to 3D-printed hydrogels. The biofabrication process begins by upcycling domestic and industrial waste into a paste-like substrate, which is sterilized and inoculated with mycelium. The paste is 3D printed into custom geometries and placed in a climate-controlled environment, allowing the mycelium to grow and bind the material into a cohesive structure. After the growth phase, the tiles are desiccated in convection ovens to halt biological activity and form lightweight, rigid panels. A second layer using a hydrogel

↑ **bioMATTERS, MYCO-CLAY, 2024**

3D-printed mycelium and clay vessel. This research explores biohybrids, various combinations of natural, bioreceptive materials and living systems. MYCO-CLAY combines mycelium-based substrates and earthenware clay bodies, yielding an aesthetic gradation between materials.

with algae-based pigments is printed into the mycelium material to add color and enhance the visibility of mycelial growth patterns within the printed material. This project contributes to the expanding field of bio-derived alternatives to synthetic colorants and to the emerging aesthetics of naturally pigmented mycelium composites.

Additional biofabrication workflows of projects at bioMATTERS include working with hybrid combinations of living organisms and earthen materials. The project Symbiont Substrate (2023) integrates 3D-printed clay modules—which are designed as bioreceptive surfaces—with automated robotic-assisted colonization of lichen cultures, composite organisms consisting of fungi and algae or cyanobacteria.[5] The ceramic components, stacked to form an architectural column, demonstrate the potential for upscaling biomaterials as modular systems using 3D printing and robotic biofabrication. This modular and adaptive approach enables incremental implementation, testing, and refinement before large-scale transformation, testing architectural scales, with greater control over form and function. The column cultivates an ongoing transformation through multispecies habitation and biodiversity, offering a vivid case of a living form with an evolving ecological aesthetic and performance. Other ongoing work with 3D-printed clay as a bioreceptive material includes MYCO-CLAY (2024), a series of vases exploring new aesthetic and material possibilities through the interplay of growing mycelium-based composites and earthenware clay bodies.[6] Here, the mycelium acts as a bio-glaze and results in a gradient transition from the living organism to the natural material.

Each of these organisms offers unique capabilities—biodegradability, low energy input, toxin-free life cycles—that contribute to environmentally sensitive fabrication. Through these systems, surfaces become active ecological interfaces, transforming into carbon-positive structures that grow, heal, and ultimately decompose naturally. The methodologies developed here reflect a cross-disciplinary approach that merges microbiology, computational design, and digital fabrication to guide and shape the growth of hybrid materialities—partially designed, partially living—which blur the boundaries between the natural and the artificial.

↓ **bioMATTERS, Symbiont Substrate (detail), 2023**

Exhibited at *rE-ordering Architecture: Making Machine and Material Kin*, Usagi Gallery, Brooklyn, New York, May 5–25, 2023, as part of NYCxDesign Festival 2023, Symbiont Substrate is a 3D-printed ceramic column that serves as a bioreceptive substrate for lichen colonization, enhancing biodiversity in urban environments. The lichen cultures are applied using an automated, robotic-assisted workflow, and custom end effector, exploring possible kinships between living organisms and machines.

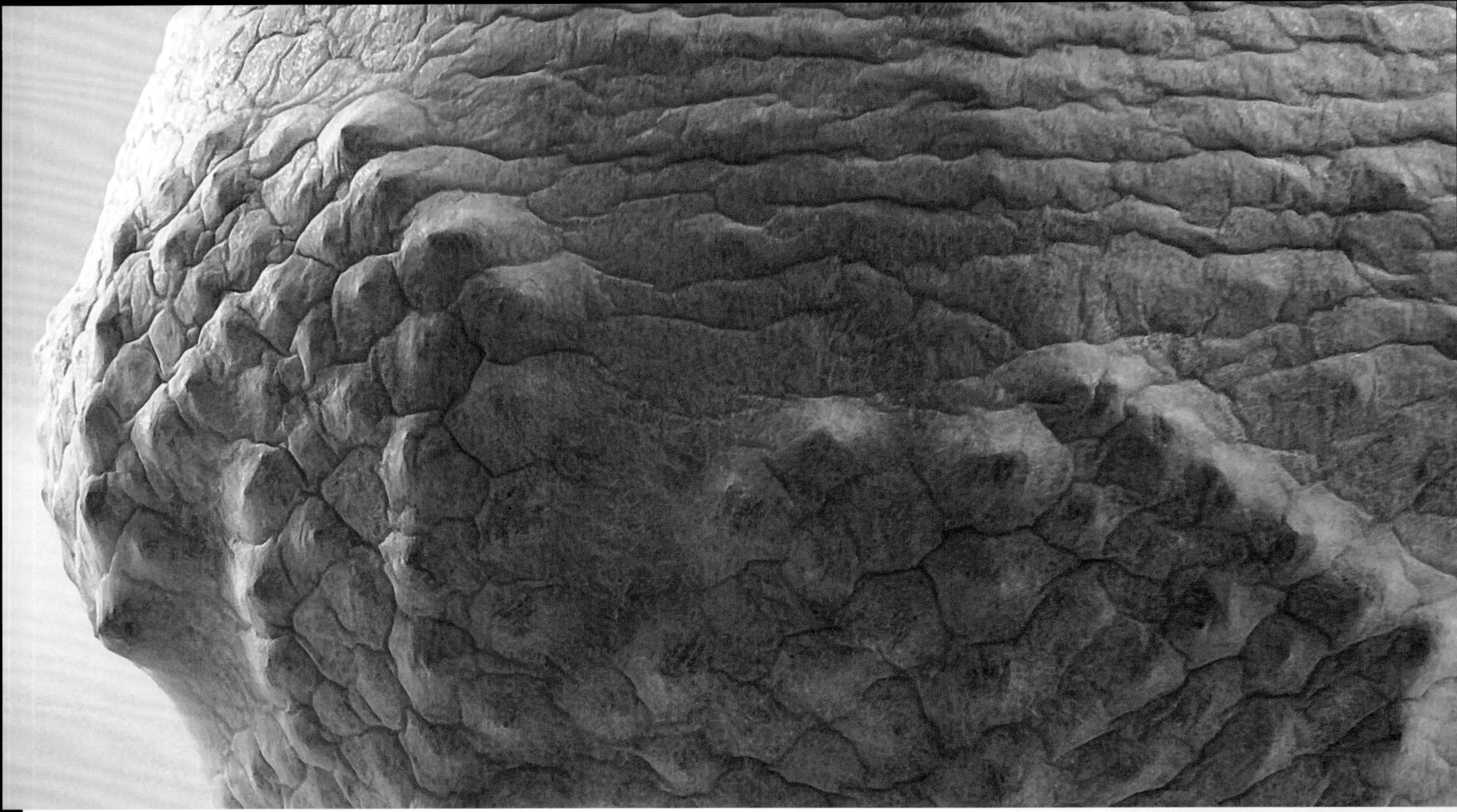

Robots and Microbes in the Age of Remediation

In 2024, bioMATTERS developed the project Hyper-accumulating Architectures: Micro Infrastructures for Mycoremediation Strategies, a work commissioned by the Museum of Architecture and Design (MAO) and Center Rog, in Ljubljana, Slovenia. The initial impetus of the project was to investigate the integration of mycoremediation techniques with conceptual design to address heavy metal soil contamination, particularly in the aftermath of the catastrophic floods that struck northern Slovenia in August 2023. These floods, described as the most extreme in recent decades, tragically resulted in the loss of human life and caused significant environmental damage, including the mobilization of heavy metals such as lead, cadmium, and zinc in the Meža Valley, a region with a long history of mining activities. Leveraging the natural capabilities of fungi and plant species known as hyperaccumulators, the research aims to create regenerative solutions, underscoring the critical role of water in soil health and ecosystem conservation. The project advocates for biomaterials and bioremediation strategies to safeguard biodiversity against the escalating impacts of climate instability.

Hyperaccumulating Architectures prototypes an integrated bioreceptive and biocomposite material composed of locally sourced materials and an organic substrate that fosters the growth of local native mycelium strains. These fungi allow for mycoremediation—the ability of fungi to absorb and isolate toxins from soil—positioning it as a potential solution for mitigating contamination in future flood-prone areas. The project materializes in seven prototype objects, 3D printed using a mix of local Slovenian dark earthenware clay and local upcycled waste as a bioreceptive substrate inoculated with regional mycelium strains.

← **bioMATTERS, 3D-printed mycelium surface texture, 2024**

Detail of a 3D-printed mycelium bowl. This research explores biodigital craft and aesthetics that conflate computational precision with naturally forming patterns and texture emerging from mycelium growth. The 3D-printed mycelium bowl series are inspired by traditional beads craft found in multiple cultures around the world, decorated with various geometric patterns and biomimetic motifs.

1 Bruno Latour, *Science in Action* (Cambridge, MA: Harvard University Press, 1987).
2 Donna Haraway, "A Cyborg Manifesto: Science, Technology, and Socialist-Feminism in the 1980s," *Socialist Review* 80, no. 15 (1985): 65–108.
3 Olivier Guillitte, "Bioreceptivity—A New Concept for Building Ecological Studies," *Science of the Total Environment* 167, nos. 1–3 (1995): 215–20.
4 Nancy Diniz and Frank Melendez, "Hybrid Bio-Based Architectural Systems: Living Organisms and Upcycled Waste Materials for 3D Printing and Robotic Deposition," in *HUMAN-CENTRIC: Proceedings of the 28th International Conference of the Association for Computer-Aided Architectural Design Research in Asia (CAADRIA) 2023*, vol. 2 (Hong Kong: Association for Computer-Aided Architectural Design Research in Asia (CAADRIA), 2023), 321–28.
5 Nancy Diniz and Frank Melendez, "Symbiont Substrate: A Bio-digital Fabrication Method for Large Scale 3D Printing of Bioreceptive Clay Modules and Automated Colonization of Lichen Communities," in *ACADIA 2023: Habits of the Anthropocene; Proceedings Two*, ed. Assia Crawford, Nancy Diniz, Richard Beckett, Jamie Vanucchi, and Marc Swackhamer (Denver: ACADIA, 2023).
6 Nancy Diniz and Frank Melendez, "MYCO-CLAY: 3D Printing Mycelium and Clay Biocomposites," in *Biodigital Intelligent Systems—Proceedings for Sociedad Iberoamericana de Gráphica Digital 2024*, November 13–15, 2024, UIC Barcelona, Spain (SIGraDi, 2024).

The Slovenian context and its unique landscapes played a major role in the form-making process of the project. Naturally sculpted formations tell a story of geological processes shaped over millennia, depicting the beauty of river and rainwater in forming the landscape, in contrast to the sudden devastation of 2023. The form-making process aimed to blur the boundaries between the natural and artificial by employing scanning data of partial segments of Slovenia caves and geological features. The datasets were then manipulated with machine learning algorithms and AI to interpret biological data from, and generate, biomorphic forms. By scanning and analyzing these intricate morphologies—layered growth, sedimentary striations, porous surfaces—the team computationally translated them into 3D models. With the use of AI and digital fabrication, the team shaped 3D-printed mycelium and clay structures, designed not only to reflect the aesthetic complexity of the landscape but to actively enhance bioreceptive properties. In so doing, niches and pockets were created as micro-climates, to effect moisture retention, microbial adhesion, and bioremediation capacity.

The concepts embedded in the projects described here investigate the theoretical implications of integrating 3D printing with living cells in the design and fabrication of objects that conflate humans, machines, and micro-organisms. These methods open new avenues for architectural speculation, marking a shift from traditional systems thinking toward living systems thinking. The projects explore how the incorporation of living materials prompts a rethinking of form, function, temporality, and material life cycles—challenging conventional notions of care, authorship, permanence, and control in design.

Additive manufacturing and computational design technologies enable new modes of biodigital craft, combining the precision of automation with the responsiveness of biological growth. Collectively, the central themes of this research reveal the aesthetic and ecological potential of biodigital hybridity, merging computational design with the generative capacities of living systems. This emerging field redefines fabrication, materiality, and the act of designing itself by embracing the dynamics of biological processes.

Working with living and biodegradable materials reorients conventional design values—conceptual, functional, temporal, and aesthetic—anticipating a future where biological rhythms and metabolic processes are not only accommodated but become integral to the creative act. Looking forward, scalable biofabrication will be enabled by accessible bioreactors, modular fermentation platforms, and robotic tools, allowing small-scale labs to grow living materials at architectural scales.

This shift toward hybrid fabrication environments—where digital and biological systems converge—will redefine how and where materials are produced. Decentralized, open-source biomanufacturing will further democratize biodesign, fostering diverse, place-based innovation and regenerative practices across global contexts. Artificial intelligence and robotics will accelerate the field by simulating biological behavior, optimizing fabrication processes and enabling adaptive, responsive systems. Together, these developments place biodigital craft at the forefront of design innovation—reshaping material culture, ecological relationships, and design agency through programmable, living systems.

Living Machines and Cultivated Architectures

Architecture may soon no longer be built but may instead be cultivated by biotechnological devices containing living materials. To maintain its living form in urban environments, biological matter must be connected to sensory systems, digital codes and feedback loops that operate at all scales—from genesis to decomposition and regeneration. These are not metaphors. These systems activate biological processes such as carbon sequestration, biodegradation, bioreception, the nutrient cycle, and so on, which are integrated into their synthetic framework. Beyond formalist and optimization approaches, this vision of design is based on metabolic principles that influence not only function, but also visual and spatial expression.

A decade ago, advanced computation was far less accessible. Today, however, digital processes are ubiquitous, spanning quantum and bioinformatic systems. What once seemed speculative has become the norm. In workshops and laboratories around the world, architects, designers, and artists are collaborating with scientists to integrate metabolically active material systems into synthetic structures that evolve with their environment. Yet one crucial step remains: transferring material innovation from the lab to the building site, which is still constrained by industrial, political, and economic systems. While large-scale construction has introduced material innovation through regulatory standards, carbon protocols, and sustainability measures, often presenting the innovation as part of a "greening" strategy, ecological conditions cannot be reduced to such fixed parameters; they emerge from complex, adaptive systems. Aesthetics, as a link between nature and culture, offers a more intuitive way of integrating these innovations into the urban landscape.

↰ **Maria Kuptsova with contributions from Sergey Kostyrko, Artem Konevskikh, Korbinian Enzinger, Gleb Andreev, and Marina Muzyka, ARBOR.Silva, fungal interface, Innsbruck, Austria, 2025**

The bio-sensorial interface of ARBOR. Silva connects mycelial networks to digital algorithms with embedded bioelectric sensors. This data is translated in real time into a digital interface, which not only monitors but also memorizes and compares the data continuously, enabling the structure to function as both habitat and interface.

Sensing **Aesthetics**

In the eighteenth century, the philosopher Alexander Gottlieb Baumgarten defined aesthetics as a form of knowledge rooted in sensory experience. For him, perception reveals deeper, intangible aspects of reality, making the relationship between the exterior and the interior an essential path to understanding. Gregory Bateson[1] and Frei Otto[2] saw patterns as bridges between nature and aesthetics—a principle now amplified by real-time biosensing. Today, technologies allow us to decode biological signals, to track the impulses of plants, the activity of fungi, and the metabolism of forests, and to integrate this data into computer design processes. What was once an iterative approach to physical modeling is now becoming a conversation between the environment and the machine.

From **Form** to Intelligence

The ARBOR project, a "cyborganic"[3] timber architecture embedded with the intelligence of a living tree, aims to find a method for bringing the complexity embedded in the natural arboreal system into synthetic form. It was initiated by Maria Kuptsova during her PhD at the Synthetic Landscape Lab at the University of Innsbruck, Austria, in 2020, under the supervision of Professor Claudia Pasquero, and evolved into a long-term research initiative—ARBOR.art[4]—which developed over five years through four major iterations. The project explores how biological memory can be extracted, analyzed, and re-expressed through digital computation and additive manufacturing.

The word "arbor," meaning tree, derives from Old Latin *arbōs* or Proto-Italic *arðōs*, and evokes forest ecosystems—complex metabolic networks of multispecies relationships and material exchanges. Anthropologist Eduardo Kohn sees forests as more than collections of plants and animals[5]—they are communicative ecosystems shaped by interactions among human and nonhuman beings. Forests operate as both physical and cultural spaces, where organisms engage through vocalizations, gestures, and chemical signals. These exchanges are not linear but instead form complex dialogues among species.

ARBOR reconceptualizes an arboreal system as a new kind of artificially living object grown by intelligent technologies. It challenges the conventional understanding of design and habitation, proposing a shift toward architecture as a metabolic process engaging both human and nonhuman forms of life—microorganisms, fungi, insects, and plants. It is a regenerative system in which the entire organism is alive; but this kind of living encompasses both biological and technological forms. ARBOR creates a biotechnological material cycle with the goal of imparting biological complexity to an inorganic object. Crucially, the lifespans of this new kind of architectural object can enhance carbon storage, remetabolize waste, and biodegrade.

→ **Maria Kuptsova with contributions from Korbinian Enzinger and Artem Konevskikh, assisted by Beyza Nur Armagan, ARBOR.Pilae genesis, 19th Venice Architecture Biennale, Venice, Italy, 2025**

A custom machine learning-based design workflow was established to translate biological memory into volumetric forms. Point cloud data from microscopic timber scans is converted into voxelized scaffolds, where each voxel encodes density, porosity, and hydrophilic behavior—rendering wood's internal logic as a printable, synthetic architecture.

Encoded Memory

Wood is a dynamic biological archive. Its vessels, fibers, rings, and cells record climatic conditions, hydrological changes, and temporal rhythms. The project-specific design workflow aims at studying these patterns through an extensive dataset of botanical characteristics and teaching the machine learning algorithm to understand them and replicate them in a synthetic form.

ARBOR focuses on local species, analyzing their microscopic structures through three anatomical sections and magnifications (×40, ×200, ×400). The first iterations of the project used generative adversarial networks trained on wood anatomy microscopic image datasets. Although effective, early approaches struggled to scale up to very detailed images, and subsequent phases therefore adopted more advanced diffusion-based machine learning algorithms. These models now translate material qualities—such as grain direction, porosity, or surface tone—into visual outputs guided by natural language prompts, allowing for controlled variations aligned with architectural intent, rather than with random outputs. Through this computer lens, the biological memory of wood becomes a generative design tool, the results not aesthetic simulations but material expressions of biological logic.

As a next step, ARBOR is developing a method for translating the complexity of these point clouds—dense 3D maps of spatial coordinates—into volumetric models, preserving their resolution while allowing further processing and fabrication. The point clouds are optimized to form voxelized scaffolds, grids of tiny volumetric pixels that structure the data for fabrication, with each voxel containing volume and density data. Custom resculpting techniques convert this data into printable formats, accommodating robotic tools and extrusion feedback.

← **Maria Kuptsova with contributions from Korbinian Enzinger, ARBOR, water-based bio-printing, University of Innsbruck, Austria, 2025**

ARBOR explores a water-based material system composed of recycled wood powder and biopolymers such as agar or chitosan, using waste from the local timber industry in Tirol, Austria, to propose wood waste as an abundant, renewable resource for 3D printing. A custom end effector regulates deposition by balancing viscosity, extrusion pressure, and flow rate, with a 40 × 40 cm unit serving as a prototype for scalable fabrication. In further tests, fungal spores (*Pleurotus djamor*) embedded in the paste grow through the structure—enhancing strength, supporting microclimate formation, and contributing to carbon capture.

Recent iterations of the project have expanded in scale, testing new material systems through continuous path robotic printing and large-scale additive manufacturing using WASP printers. The process uses industrial sawdust, transformed into two composite types: a fibrous mix (40 percent recycled wood powder and 60 percent PLA, a biodegradable thermoplastic) and a biodegradable pulp of wood waste combined with chitosan, a natural polymer extracted from shellfish. The feedback between fabrication and coding was key to refining print quality.

Water-based 3D printing revealed challenges with mechanical consistency due to regional substrate variations. ARBOR sourced local sawdust—typically considered waste—and repurposed it as reusable print material. A custom end effector—the specialized tool mounted at the end of a robotic arm—manages deposition, with extrusion parameters fine-tuned for scalability. To enhance structural and metabolic function, ARBOR integrates fungal mycelium. Pink oyster mushroom (*Pleurotus djamor*) spores and nutrients are embedded into the paste, sprouting through the structure to compact it, generate microclimates, and boost carbon capture. Fungal and wood species are selected for ecological compatibility, though local agricultural conditions influence their interaction. This specificity complicates standardization but allows adaptation to environmental and regulatory contexts.

Fibrous composite experiments demonstrate circularity by replacing nearly half of the material mass with timber waste. Coatings like PURE.TECH absorb carbon dioxide and nitrogen oxides (NOx), creating a synthetic metabolism. The resulting materials maintain organic tactility and scent, biodegrade, and nourish new biological life as they decompose.

→ **Maria Kuptsova in partnership with LaMáquina, ARBOR. Pilae, cultivated matter, Barcelona, Spain, 2025**

A pallet-based wood composite demonstrates circularity by partially replacing the inorganic mass with timber waste. Coatings such as PURE.TECH absorb CO_2 and NOx, generating a synthetic metabolism. The resulting materials retain organic tactility and scent, biodegrade naturally, and nourish new biological life as they decompose.

ARBOR.Pilae: The **Column** as Archive

Presented at the 19th Venice Architecture Biennale in 2025, curated by Carlo Ratti, ARBOR.Pilae reflects Venice's historic relationship with wood. The city's foundations rest on oak piles sourced from the Cansiglio Forest—structures that have withstood the effects of water for centuries. Referencing this legacy, ARBOR.Pilae explores new forms of bio-artificial architecture rooted in the structural intelligence of biological systems. The printed structure evokes a tree, the architectural archetype of the column. Exhibited in the Corderie dell'Arsenale, its proportions reflect the neighboring brick columns, forging a spatial and morphological dialogue. Originally conceived as a plinth, the base has evolved into a root system scanned from a local oak tree, creating a foundation informed by the complexity of the tree root system. The internal morphologies echo the distribution of biological materials, transforming the column into a stratified botanical archive that reveals each species through the shape of the section.

→ Maria Kuptsova, ARBOR.Pilae, exhibited in the Corderie dell'Arsenale at the 19th Venice Architecture Biennale, Venice, Italy, 2025

Rather than decorating or imitating arboricultural motifs, ARBOR studies and translates the anatomical data of timber into vertical geometries that embody both internal and external structures, establishing a new typology for architectural objects informed by the structural intelligence of biological systems. ARBOR.Pilae reflects the neighboring brick columns of the Corderie dell'Arsenale, forging a spatial and morphological dialogue that reinterprets architectural typology through the arboricultural logics of local species.

→ **Maria Kuptsova in partnership with LaMáquina, ARBOR. Pilae, metabolic energy, Barcelona, Spain, 2025**

ARBOR.Pilae follows principles of growth and metabolic function, with convoluted surfaces transitioning from vascular interiors to bark-like envelopes. This continuous movement between internal fibers and external topologies dissolves the boundary between structure and ornament.

↓ **Maria Kuptsova in partnership with LaMáquina, ARBOR. Pilae, pre-assembly test, Barcelona, Spain, 2025**

Custom techniques for algorithmic resculpting tailored to specific robotic and 3D-printing parameters optimize the transformation of anatomical data into printable formats, enhancing the feedback loop between design and robotic fabrication, which allows a three-meter high-resolution synthetic living sculpture to grow.

Rather than simply decorating or imitating arborical motifs—as traditional architecture has done through its system of orders—ARBOR translates anatomical data into vertical geometries that fuse both internal and external forms. The resulting morphology follows the logic of growth and metabolic function while retaining a high degree of complexity. Surfaces become complex and convoluted, moving fluidly from internal vessels and fibrous networks to high-performance, bark-like topologies, then back inwards, creating continuities between structure and ornament. The sculpture renders ecological memory in synthetic form.

ARBOR.Silva: Fungal **Computing**

ARBOR.Silva, developed between January and March 2025, extends the "cyborganic" paradigm by investigating the bioreceptive potential of wood-based architectures integrated with living fungal intelligence. This iteration of the project explores how high-resolution synthetic structures can not only host but also actively engage with fungal life, functioning simultaneously as architectural material and biological interface.

Often called the "Wood Wide Web," the mycological communication network is capable of transmitting information between different arboreal species, linking several plants in an ecosystem. Through this underground web, trees and plants can exchange water, nitrogen, carbon, and biochemical signals, enabling communication between species and the sharing of resources on the forest floor.

Informed by the mutualistic symbiosis between trees and fungi in forest ecosystems, ARBOR.Silva integrates spores of *Pleurotus djamor* into a printed wood matrix enriched with organic nutrients. The bioreceptivity of the project operates on several levels. On the outside, its structure invites colonization, facilitating visual and ecological changes over time. On the inside, its capillary geometries allow the exchange of moisture and gases, creating micro-environments conducive to fungal growth. The mycelium spores, distributed in the internal cavities of the structure, germinate into mycelium that colonizes the surface and substructure. As the living fungal network expands, it acts as a bioindicator: Its growth patterns change in response to changes in temperature, humidity, light, and air quality, providing a living diagnostic layer that reveals the surrounding environmental conditions.

ARBOR.Silva draws on scientific advances in fungal cognition and unconventional computing. Research by Andrew Adamatzky demonstrates that mycelial networks are capable of transmitting electrochemical signals that mimic primitive information processing.[6] ARBOR.Silva builds on this knowledge by integrating conductive sensors into the structure that detect the electrical fluctuations generated by the mycelium. These biosignals—rich in temporal and spatial complexity—are recorded and processed using machine learning techniques to interpret the physiological state of the fungi in relation to environmental variables. In parallel, diffusion models reconstruct spatial distributions, mapping fungal responses through structure topology. The resulting data is not just recorded but also translated into a real-time interface, which allows analysis of the large complexity of data in time.

A system of bioelectric sensors weaves the sculptural components into technological branches. Each wire acts as a vein, transmitting signals between the biological organism and digital algorithms. These exposed elements invite reflection on the aesthetics of technological systems, drawing a critical comparison between synthetic wiring and the organic fibers of mycelium. Scaling up this system and envisioning the inhabitation of such architectures would require a new kind of maintenance—a type of "gardening"[7] attuned to biological rhythms through engagement with biodigital codes and synthetic interfaces.

1 Gregory Bateson, *Steps to an Ecology of Mind: Collected Essays in Anthropology, Psychiatry, and Epistemology* (Chicago: University of Chicago Press, 2000), 105.
2 Frei Otto, *Occupying and Connecting: Thoughts on Territories and Spheres of Influence with Particular Reference to Human Settlement* (Stuttgart: Edition Axel Menges, 2009).
3 Maria Kuptsova, "Cyborganic Living," in "Machine Hallucinations: Architecture and Artificial Intelligence," ed. Matias del Campo and Neil Leach, *Architectural Design* 92, no. 3 (2022): 116–21.
4 "Arbor.art: The Cyberorganic Living Objects, Bio-Artificially Grown by the Means of Intelligent Technologies," Arbor, accessed August 19, 2025, https://arbor/art/.
5 Eduardo Kohn, *How Forests Think: Toward an Anthropology Beyond the Human* (Berkeley: University of California Press, 2013), 18–25.
6 Andrew Adamatzky, "Language of Fungi Derived from their Electrical Spiking Activity," *Royal Society Open Science* 9, no. 4 (2022).
7 Gilles Deleuze and Félix Guattari, *A Thousand Plateaus: Capitalism and Schizophrenia*, trans. Brian Massumi (Minneapolis: University of Minnesota Press, 1987).

← **Maria Kuptsova with contributions from Sergey Kostyrko, Korbinian Enzinger, Artem Konevskikh, Marina Muzyka, and Gleb Andreev, ARBOR. Silva, biodigital living scaffold, Innsbruck, Austria, 2025**

The bioreceptivity of ARBOR.Silva operates on multiple levels. Its outer structure encourages colonization and visual transformation, while internal capillary geometries enable moisture and gas exchange, creating micro-environments for fungal life. The wooden-fungi substratum becomes a living scaffold, where expanding mycelial networks respond to environmental changes, forming a dynamic, bio-indicative surface.

Living Intelligence of **Architecture**

Architecture today increasingly emerges from recursive exchanges between biological, computational, and material systems. Machine learning algorithms mediate between living systems and fabrication, enabling forms that behave less as static enclosures and more as evolving ecosystems. Inhabited by bacteria, fungi, insects, plants, and humans, these environments evolve through biodigital design methods that fuse biological intelligence with technological processes.

The next crucial step is to translate this material innovation into built environments where performance is balanced with aesthetic experience. As architecture becomes a living entity, its aesthetic qualities arise from the interplay of organisms, algorithms, and matter—reshaping how we design and inhabit the world.

PROBIOTIC TECTONICS

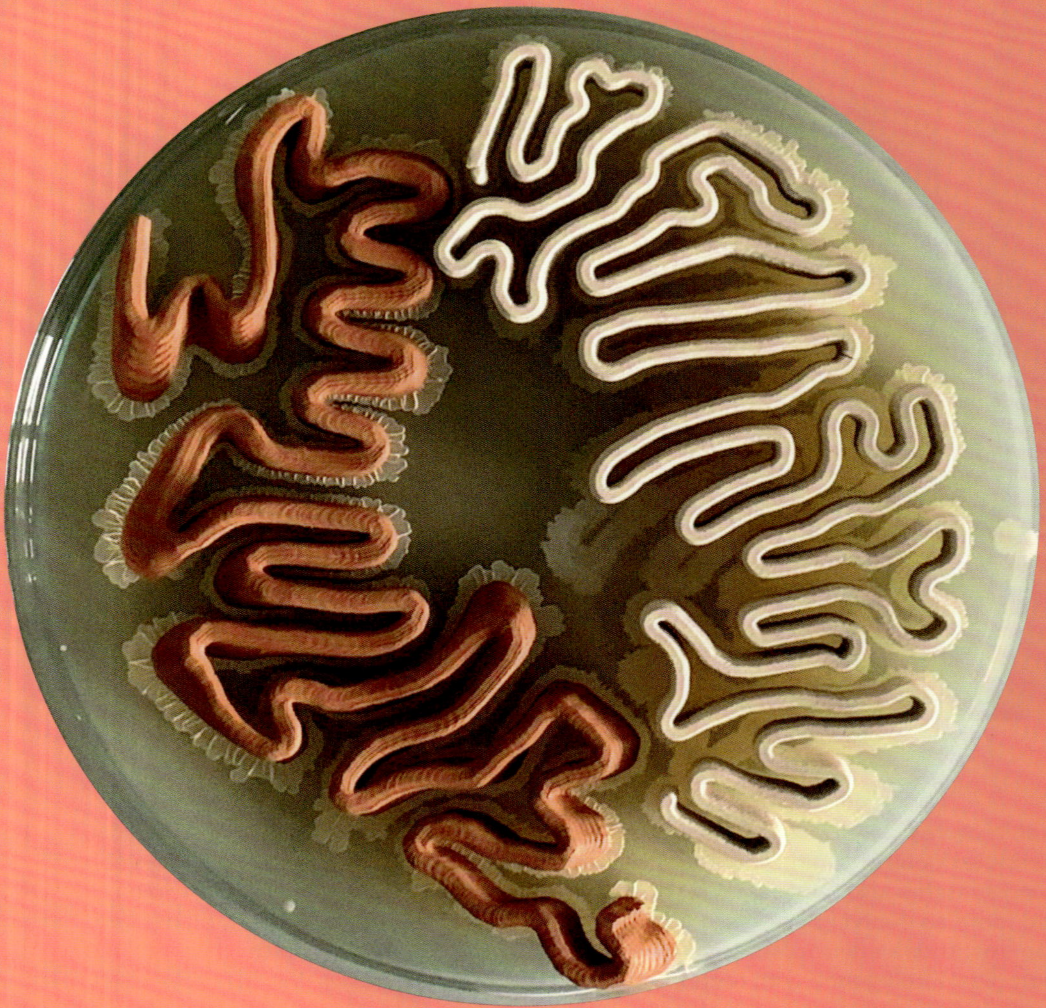

Richard Beckett

DESIGNING FOR MICROBIOMES

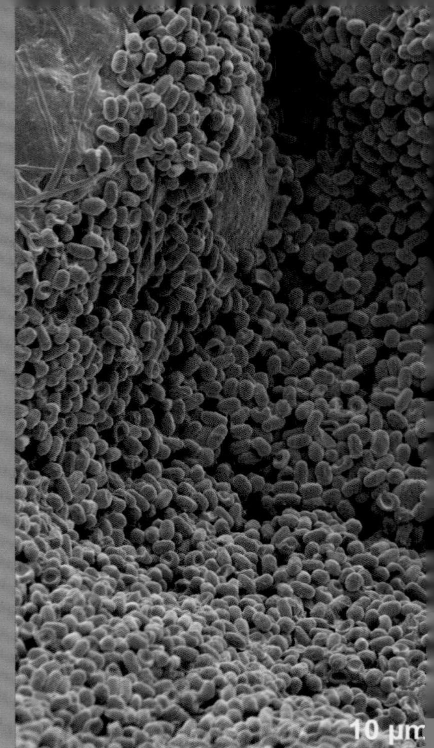

Microbiome-Inspired Design

Designing for the microbiome offers new lines of investigation for the field of biodesign in architecture. The microbiome defines the communities of microorganisms (including fungi, bacteria, and viruses) that exist in a particular environment, including buildings and bodies. This paradigm emerges amid a growing recognition that humans are not isolated biological entities but complex *holobionts*—ecological assemblages composed of human and microbial life. This challenges the enduring legacy of modernist design, which imagined the body as sealed, sterile, and separate from its environment. In contrast, emerging microbiome science reveals that our health, cognition, and behavior are deeply entangled with the microbial communities that live on, in, and around us.[1] Contemporary microbial pathologies, including diseases of "missing microbes,"[2] appear to be related to the lack of exposure to microbial diversity in contemporary cities. This calls for a shift in architectural thinking: from buildings as barriers to infection, to buildings as living interfaces that foster microbial diversity, promote interspecies cohabitation, and facilitate beneficial microbial exposures.

Current paradigms in building health still largely rely on the outdated assumption that fewer microbes equate to healthier environments. These ideas are rooted in nineteenth- and twentieth-century modernism, when architecture was imagined as a hygienic cure for industrial urban disease. As the architectural historians and theorists Beatriz Colomina and Mark Wigley have stated, modern architecture "explicitly defined itself in opposition to bacteria."[3] Such mindsets shaped the materiality and tectonics of architecture, favoring smooth, sealed materials, and minimal surface articulation—flat white walls and straight-line forms. Since the mid-twentieth century, these aesthetics have aligned with the goals of energy efficiency and climate control, resulting in increasingly sealed buildings. Indoor environments became isolated from the natural world, limiting exposure to the microbial diversity essential for human health.

Reframing Architecture-Microbe Relationships

The concept of the human holobiont offers a radical shift in how we understand not only the body but also architecture. No longer separate from the ecosystems around us, bodies are understood as deeply porous—biologically dependent on the microbial diversity that surrounds them. Microbes are not contaminants but coevolutionary "old friends" essential for digestion, immunity, and cognition.[4]

As our primary habitat, buildings must also be reimagined as porous, symbiotic environments that cultivate microbial diversity rather than suppress it. This requires designing architectural porosity across multiple scales: at the microscale of material properties, the mesoscale of textures and articulations on surfaces and envelopes, and the macroscale of building form, tectonics, and spatial programming. Probiotic design reframes buildings as ecologically active "immune systems" where form, space, and materiality modulate indoor microbiomes to benefit both environmental and human health. It proposes new material and spatial ethics where biocolonization, exposure, porosity, ageing, decay, and erosion are no longer seen as threats but as opportunities for microbial interaction, care, and cohabitation.

In doing so, probiotic design shifts thinking, from sterility and control toward multispecies entanglements and long-term ecological participation. At the Beckett Lab, part of the Bartlett School of Architecture at University College London (UCL), ongoing interdisciplinary research explores these ideas through the development of biologically active materials, surfaces, and spatial systems. This work asks: How can we rewild interiors to promote microbial richness? How can architecture foster healthy microbial exchanges? And how might tectonics become adaptive, eroding systems that evolve over time?

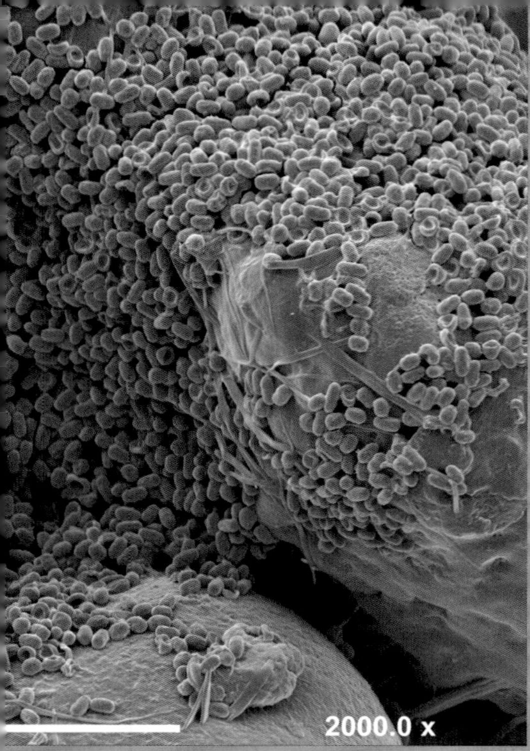

↰ **Beckett Lab, Developing Probiotic Interventions research project, Bartlett School of Architecture, University College London, 2023–26**

Two different bioreceptive clay materials inoculated with the same *Bacillus subtilis* strains. As part of the material development research stage, these experiments make visible how the physical and chemical properties of materials create "niches" that can shape microbial agency, here exhibiting different biofilm formations, resulting in different performativity.

↑ SEM image of *B. subtilis* colonization in bioreceptive ceramic matrix, with cells undergoing form transition toward spores, an evolutionary strategy that supports survival. In this spore state the microbes can survive for extended periods of time indoors without the need for moisture or nutrients.

Materials as Ecological Niches

Developing Probiotic Interventions is a project at the Beckett Lab, funded by the UK's Engineering and Physical Sciences Research Council (EPSRC). Building on the concept of bioreceptivity[5] and bioreceptive design[6]—where materials are engineered to support intentional biocolonization—this research extends the focus from exterior façades and photosynthetic organisms to the complexities of indoor environments and the human body. This shift brings the indoor microbiome into focus, entangling biologically active materials, microbial agencies, and the architectural spaces where bodies dwell. Probiotic materials in this context serve dual roles: firstly, as protective agents where benign bacteria outcompete harmful pathogens, and secondly, as ecological enrichment devices that introduce microbial diversity associated with natural environments—what the medical microbiologist Graham Rook terms our "old friends."[7]

Indoor spaces, however, are typically dry, nutrient-poor, and highly selective, and this determines not only which microbes can survive but also *how* they behave. Microbes resist linear, predictable behaviors, and are often described as "unruly." Thus, designing with microbes demands a move from control to relational engagement. This means going beyond simply hosting microbes, to designing materials in ways that influence microbial functionalities—shaping how microbes inhabit, adapt, and interact within material matrices. For example, materials can be engineered to promote spore formation—a resilient microbial state that supports long-term survival indoors without the need for moisture or nutrients.

Probiotic materials can thus be viewed as ecological niches, aligning with psychologist James J. Gibson's theory of affordances: opportunities for action offered by an environment to an organism.[8] Rather than seeing materials as simply passive hosts, this approach reimagines them as active ecologies that support microbial communities and multiple functionalities. Designing for probiotic agency involves creating material affordances that nurture beneficial microbial behaviors, something achieved by experimenting directly with microbes and adjusting materials in response to microbial feedback.

Experiments embedding *Bacillus subtilis* into various bioreceptive clay matrices make visible how microbial agency varies according to the material context and the environment they are placed into, reinforcing the intertwined nature of these interactions. Probiotic performativity, then, is a relational process between material, microbe, and environment—rather than a fixed property. By designing for these relationships, probiotic materials can support protective affordances, enabling functions such as the production of metabolites that aid pathogen inhibition.

Extending the metaphor of buildings as immune systems, probiotic materials can be designed to respond, dynamically, to anticipated microbiome and environmental changes. Environmentally informed material strategies can afford bacteria's ability to form biofilms or transition to spores, which can then reactivate when triggered by environmental cues, contributing to more adaptive, resilient indoor microbiomes.

↑ **Beckett Lab, Probiotic Symbiosis, Bartlett School of Architecture, University College London, 2022**

Consisting of five interlocking components, the probiotic surface prototype is inoculated with soil microbes cultivated from rural soil samples. Environmental-driven design and fabrication processes inform geometrical conditions of porosity, maximizing the surface area of probiotic material exposure at the air-surface interface. During an intervention study, the prototype was shown to directly increase the microbial diversity of an office space.

Modulating Indoor Microbiomes

The contemporary understanding of the immune system is one that *modulates* beneficial microbes rather than only defending against invasive pathogens. When applied to architecture, this informs the design of indoor microbiomes that do more than just resist contamination. Probiotic materials can actively contribute to and shape the indoor microbiome, serving as sources of beneficial diversity for rewilding indoor microbiomes with microbial diversity, in ways that are healthy for human occupants.

Shifting upwards in scale from materials, the design of probiotic surfaces in the Beckett Lab explores the potential to rewild buildings and bodies with beneficial microbial diversity. This work engages the spatial relationships between building surfaces, microbes, and the immaterial conditions of indoor spaces through what are called mesoscale design parameters. Extending Gibson's ecological theory, surfaces are conceived as porous interfaces. In material science, interfaces mediate behaviors such as reflection or absorption; in this case, the biological interface mediates microbial exchanges between buildings and bodies. This incorporates geometrical conditions—including surface articulations, textures, and tectonic conditions—that shape the interfacial conditions between probiotic surfaces and indoor microbiomes.

Computational design and digital fabrication methods are used by the Beckett Lab to explore working prototypes of biologically active tiles, wall panels, and surface components. These are then tested as intervention studies in real-world environments, using contemporary techniques of metagenomic sequencing and environmental monitoring, in order to understand and quantify their effect on the indoor and human microbiome.

Microbial Shedding and the Productive Potential of Erosion

Designing for the intentional transfer of microbes from probiotic surfaces to buildings and bodies reframes microbial transmission as an opportunity for symbiotic exchange and for new modes of cohabitation between microbes, architecture, and humans. This thinking subverts the phenomenon of surface and airborne microbial transmissions made visible during the COVID pandemic, toward something much more entangled and beneficial for health.

Central to this is the concept of "microbial shedding," where biologically active, probiotic surfaces become productive agents in shaping indoor microbial ecologies. In the Beckett Lab's project Probiotic Symbiosis (2022), indoor airflow simulations were used to model geometric surface components, which could shed their probiotic microbes as a result of airflow turbulence at the air/surface interface.

Similar ideas were explored in the project TerraSolum (2023), whereby the living surface was designed to erode and shed its own material, actively transferring its beneficial microbes into the air and onto nearby surfaces. Challenging modern ideals of cleanliness, the project used "dirt," in the form of biologically rich soil materials containing diverse forest-floor microbiota, which were integrated into the rough, articulated surface condition.

This approach coincides with biodesign thinking, diverging from Western ideals of permanence, newness, and impermeability—values rooted in modernist "antibiotic" paradigms. The material mix used in TerraSolum was deliberately fragile, with various material compositions explored alongside geometrical forms to limit post-fabrication resilience and facilitate time-based erosion during everyday use.

Designing for shedding and erosion must acknowledge the challenges of maintenance. However, by aligning with ecological thinking, such concepts do allow for adaptability. Indoor microbiomes are inherently dynamic, shifting with seasons, building use, and human activity. In this way, erosion and the maintenance that follows become a productive aspect, enabling the building to respond quickly to inevitable changes and various next states. This not only facilitates environmental resilience but also permits performative and aesthetic variation.

↓ **Beckett Lab, TerraSolum, London Design Biennale, Somerset House, London, 2025**

Probiotic concrete wall panels integrated with micro-biodiverse soil material derived from forest floors in Finland (material supplied by Uute Scientific). The panels were first exhibited at the *Small Spaces in the City: Rethinking Inside the Box* exhibition at Roca London Gallery, London, 2023. The living panels formed part of a live, indoor intervention study to measure the effect of beneficial microbial exposures on the human immune system.

↘ **Beckett Lab, TerraSolum panel prototype, Bartlett School of Architecture, University College London, 2025**

Geometric depressions and protrusions combine with fragile material mixes and soil densities to reference aesthetics of ageing, ancientness, and decay. These conditions increasingly manifest over time during use, when the material erosion "sheds" its beneficial microbes, serving as a source of "old friends" for the holobiont body and the indoor microbiome.

Holobiont Entanglements

Concepts of ageing and erosion also allow for new human-microbial interactions through notions of care, and "tending to" living surfaces. This thinking engages the concept of the holobiont, where inhabiting with and tending to microbes further facilitates beneficial microbial *entanglements*. These entanglements occur through both direct contact (e.g., touch) and indirect pathways (e.g., respiration), allowing microbes from the air and surfaces to integrate with the human microbiome on the skin, in nasal passages, and within the gut. Such interactions can reconfigure the holobiont, linking architectural porosity to the body's porous and coevolutionary nature.

At the architectural scale of building tectonics, establishing the building fabric as a permeable, bioreceptive interface connects and modulates the outdoor environmental microbiome of building surfaces with indoor microbiomes and holobiont occupants. A prototype proof of concept, Probiotic Tectonics (2022), was developed to explore the viability of large-scale probiotic building elements.

The outer surface of the prototype is designed to support moss and algal colonies, which contain within them—and host—their own diverse environmental microbiota. Behind this surface, several layers of bioreceptive materials are embedded within a semi-structural, 3D-printed concrete lattice. These layers vary in their pore sizes, creating a porosity gradient that facilitates the movement of environmental microbiota from the outer to the inner surface.

Ongoing microbiome studies are focused on examining the community composition on the inner surface, tracking how it evolves over time, and assessing the care requirements of such probiotic strategies.

← **Beckett Lab with Incremental 3D, Probiotic Tectonics, prototype, 2022**

A 1:1 scale, probiotic building element. Envisaged as part of a porous building interface, the element serves as a host for multiple ecological communities throughout its volume. The work explores how digital biofabrication approaches can facilitate outdoor-indoor movement of environmental microbial diversity, as a strategy to rewild indoor environments with beneficial microbes.

← Exterior-facing surfaces host diverse environmental microbial communities as part of bryophyte colonization, providing a resilient, low-maintenance habitat, well suited for growth in urban environments. As these microbial communities are transported inwards through the interface layers, indoor-facing surfaces are monitored and tended to shape desired microbiome functionalities.

↙ Multiple bioreceptive layers with graded porosity were embedded into the element during the robotic fabrication process, infilling between the semi-structural pattern layers.

Future Research

Probiotic design offers a novel area of research within architectural biodesign, focusing on the integration of beneficial microbes into buildings to create healthier built environments. By supporting microbial diversity, probiotic design may help protect occupants from harmful pathogens, offering a form of biological resilience that could be vital in preventing future pandemics. It also reframes the relationship between architecture and the human body, particularly the holobiont body and its associated microbial communities. While the research incorporates laboratory methods, its future potential lies in architectural applications beyond the lab. Achieving this will require intervention-based, real-world studies that assess microbial and immunological responses. Architects must collaborate with microbiologists, ecologists, and immunologists to interpret metagenomic data and track immune system changes. These interdisciplinary efforts are key to understanding how architectural design can support human-microbe interactions and promote immune health.

1 National Academies of Sciences, Engineering, and Medicine, *Microbiomes of the Built Environment: A Research Agenda for Indoor Microbiology, Human Health, and Buildings* (Washington, DC: National Academies Press, 2017).

2 Martin Blaser, *Missing Microbes: How Killing Bacteria Creates Modern Plagues* (London, UK: Oneworld, 2014), 288.

3 Beatriz Colomina and Mark Wigley, "The Bacterial Clients of Modern Architecture," *Docomomo* 62 (August 2020): 7.

4 Graham A. W. Rook and Sally F. Bloomfield, "Microbial Exposures That Establish Immunoregulation Are Compatible with Targeted Hygiene," *Journal of Allergy and Clinical Immunology* 148, no. 1 (2021): 33–39.

5 Olivier Guillitte, "Bioreceptivity: A New Concept for Building Ecology Studies," *Science of the Total Environment* 167, no. 1 (1995): 215–20.

6 Marcos Cruz and Richard Beckett, "Bioreceptive Design: A Novel Approach to Biodigital Materiality," *Architectural Research Quarterly* 20, no. 1 (2016): 51–64.

7 Graham A. W. Rook, Christopher A. Lowry, and Charles L. Raison, "Microbial 'Old Friends,' Immunoregulation and Stress Resilience," *Evolution, Medicine, and Public Health* 2013, no. 1 (2013): 46–64.

8 James J. Gibson, *The Ecological Approach to Visual Perception*, classic ed. (New York: Psychology Press, 2014).

Built Ecologies— **Metabolistic Resource Thinking Through Biogenic Materials**

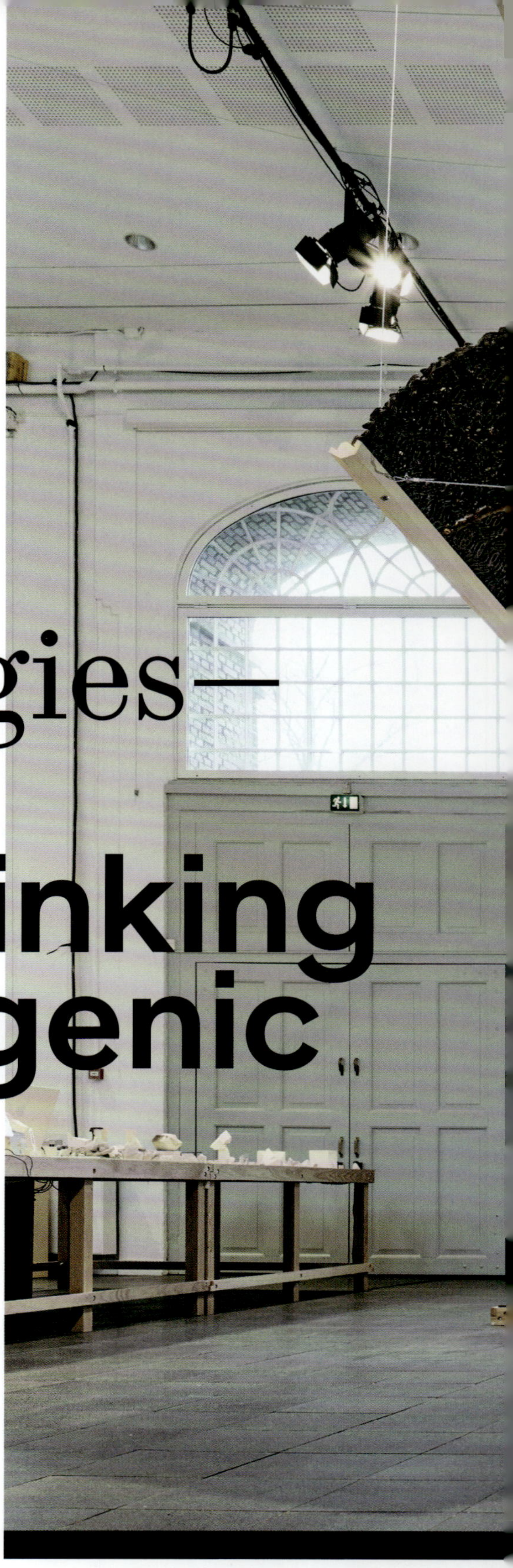

Mette Ramsgaard Thomsen and Paul Nicholas

Biogenic materials are fundamentally cyclical. Whether associated with heritage-driven approaches or high-tech innovations, they are formed by cycles of growth, deployment, and degradation. As instantiations of biomass, they flow through a global steady-state system in distinct temporal cycles ranging from those of seconds and minutes in the lives of cells and bacteria, to those of centuries and millennia in the lifespans of forests and ecosystems. It is this inherent circularity that brings forth our perception of biogenic materials as renewable.

However, many procurement processes of growing and harvesting are far from regenerative. The current focus on timber construction demonstrates this clearly. Timber is understood to be one of the few carbon-neutral building materials, yet the increased demand for construction timber relies on the extractive methods of large-scale logging and on monoculture practices. Despite sustainable harvesting efforts like selective harvesting and replanting, these practices still exacerbate biodiversity loss and forest degradation, leading to soil erosion and the disruption of water cycles while fundamentally disrupting the forests' ability to preserve carbon. At its core, the current sustainability discourse in architecture tends to overlook the complex, relational nature of resources and ecosystems, as well as the human communities that participate in them. The prevailing conceptualization of resource renewability exacerbates this contradiction. Contemporary circular design frameworks favor biocomposting and energy production as desired end points. This limits the potential for their cascading, an integral principle of circularity enabling the sequential reuse of resources to maximize biomass availability through practices of reduction, reuse, repair, and recycling.

To address this, the perception of biogenic materials needs to be expanded, and alternative material classes need to be defined. Their unique qualities need to be activated in accordance with novel cascading logics that rethink what reparatory and remediative practice can be in a material domain where regeneration is not separate but instead deeply embedded.

Three projects are presented here, which explore and exemplify these novel practices. Building on a series of research enquiries examining biopolymer composites, these projects demonstrate how bio-based composites can support a rethinking of cascading in architecture through their ability to be reformed, adapted, and remade. Each project positions metabolistic resource thinking as an extension of circular cascading. Where metabolistic thinking is foundational to industrial ecology and circular design as a descriptor of the systematic study of material and energy flows within industrial systems and their exchange with the environment, it has in these contexts often remained mechanistic in character and deterministic in its orientation toward predefined end points. Challenging this sense of predetermination, the projects do so by envisioning a more interconnected and regenerative flow of materials that adapts to changing contexts over time. Here, malleability, repair, and ecological belonging are interconnected and mutually reinforcing practices that form the foundation of a novel regenerative architecture. This is an architecture that challenges the standardization-based logics of current biogenic resource procurement, that extends the limited palette of biogenic materials, and that appreciates the need to identify and work with local resources while recognizing their socioecological dependencies.

The three projects—Rhizerial, Repair Ecologies, and Local Resource / Collective Knowledge—are full-scale investigations that position cascading biopolymer composites in metabolistic cycles of deployment.[1] Developed at the Centre for Information Technology (CITA) at the Royal Danish Academy between 2024 and 2025, these prototypical architectures redefine our ideation of cascading and, through this, our notion of end points and lifespan in architecture. By demonstrating the practices of malleability, repair, and ecological belonging, they propose ways in which design, specification, and fabrication can be understood as iterative practices that open generative potentials within adaptation and refabrication, nested into the larger life cycles of biogenic materials.

↰ **CITA, Rhizerial, *In the Making—Artistic Methods in Architecture and Design* exhibition, Royal Danish Academy—Architecture, Design, Conservation, Copenhagen, Denmark, 2024–25**

The Rhizerial installation is a five-by-four-meter suspended vault structure created from graded biopolymer composites using locally sourced fillers and fibers sourced from agricultural and industrial ecologies.

↘ The biopolymer composite is printed robotically onto a shaped mold mounted to a turntable. Two separate material feeds are mixed at the nozzle, and locally air cooled to speed reprinting. Material grading occurs throughout the plane and section of each vault.

↘ The material grading affords performance-driven and aesthetic variation. The different recipes are used to vary the cellulose sources as well as their amounts.

Malleability, Repair, and Ecological Belonging

The three projects use biopolymer composites to examine how these materials can be formed, reformed, and adapted using additive manufacturing, specifically robotic 3D printing. The biopolymer composites incorporate biogenic materials that possess inherently ecological belonging, and they are sourced from residual material streams that are deeply connected to the local agricultural and industrial ecologies they inhabit. Rooted in metabolistic resource thinking, these materials include locally sourced eelgrass, bark, wood flour, mussel shells, and cotton from reclaimed textiles. Across the three projects, resource streams are mapped and correlated with ecological dynamics, exploring how architecture and the built environment can integrate into these systems. In this context, the idea of malleability enhances the regenerative capacity of biogenic materials, allowing them to be reconfigured and reintegrated into ecological cycles. Importantly, these cycles are not static but evolve over time, reflecting the complex interplay between material flows and the changing environments they engage with.

The Rhizerial installation (2024) examines malleability in the recipe space to allow navigation across a variety of constituent fillers and fibers.[2] Built as a suspended vault structure made of fitted panels, the structure is made from graded biopolymer composites that change across the structure as well as in its depth. The material system is understood here through gradients interfacing metabolistic resource thinking with performance and deployment. Through the design and specification of different material recipes—combining a variation of fibers from various cellulose sources—this results in a structure that changes in performance, and the aesthetics they embed are graded across the suspended ceiling structure. This sense of material malleability is brought down to the scale of the nozzle: The work's fourteen large-scale panels are robotically 3D extruded using a bespoke multi-feed extruder that continuously grades the material streams in response to performance criteria.

Biopolymer composites have high response rates to environmental changes, such as fluctuations in humidity and temperature. The case study Repair Ecologies (2025), a weather screen façade, examines how material malleability and the inherent changeability of biopolymer composites can promote a rethinking of durability and lifespan in the built environment. Using the same material systems as those in Rhizerial, the project exploits the collagen binder's thermoplastic-like properties.[3] When adding humidity and heat to printed elements, the material system becomes malleable, allowing its reshaping, adaptation, and remaking over multiple cycles. Repair Ecologies exploits these properties to ideate principles of continual construction in which reparatory practice is used to reinstate breakage and deformation in the façade's panels. Repair Ecologies consists of thirty-six panels, each printed in multiple layers with articulated drip noses that lead rainwater away from the surface. The panels are monitored, and live tracking informs inspection and diagnosis. As panels fall into disrepair, they are returned to the fabrication lab where they are further diagnosed and cleaned. They are then 3D scanned and reprinted upon—using conformal printing strategies that directly address the specificity of the defects. Edges that have shrunk are reestablished, cracks are restored, and drip noses that have swollen are rearticulated.

Repair Ecologies ideates principles of continual construction in which reparatory practice is used for extending the lifespan of building elements. In this context, lifespans are seen as dynamic constructs that can be continually modified and adapted to changing conditions. Malleability affords a continual material address in which durability becomes a function of the ability to reform and adapt the material system. In doing so, malleability is placed at both recipe and material system level. At these distinct, interconnected scales, malleability is not just a material property but a strategic framework for designing adaptable, regenerative material systems that remain flexible across different contexts and stages of their life cycle.

↙ **CITA, Repair Ecologies, exhibited at the Danish Design Museum, Copenhagen, Denmark, 2025**

Repair Ecologies is a weather screen façade made from collagen-based biopolymer composites. The biopolymer responds to changes in temperature and humidity through swelling, cracking, and fusing, creating a direct link between the material system and its environment.

↓ The repair process utilizes the biopolymer composite's capacity for self-bonding. The repair actions are grounded in underlying technological processes of registration, diagnosis, and design response.

→ **CITA, Local Resource / Collective Knowledge, 19th International Architecture Exhibition, *Intelligens. Natural. Artificial. Collective*, Venice Architecture Biennale, Venice, Italy, 2025**

This installation was undertaken as a collaboration with architects 3XN and GXN and creates three biogenic façades clad in panels woven from palm fronds, grown via a bacterial process of biocementation, and cast and milled from blue biopolymer composites.

Local Resource / Collective Knowledge (2025) expands these concerns to a territorial scale.[4] The overall installation consists of three biogenic façades exploring different technological platforms, ranging from palm frond weaving, the growth of biocement, and the casting and milling of blue biopolymer composites. Created in collaboration with practice 3XN and their associated research group GXN, the project examines how concepts of remediation and resilience can become an embedded part of architectural design. The installation consists of three façades addressing three different locales, developing means of engaging with local resources and skills. In the biopolymer façade, developed by CITA, an alginate is used as a binder mixed with differentiated species of blue and white mussel shell waste as well as eelgrass fibers. The tiles are cast in formed molds and post-process milled. Each tile combines multiple material recipes, adding to the aesthetic and expression of the overall wall, which is further articulated by variations to the thickness and milling patterns of each tile.

By examining the making of architectural elements from aquatic biocomposites, the project explores how architectural material practice can connect to and participate in the socioecological practices of mussel farming. Sourced from the Po River Delta, the shells of the Mediterranean mussel *Mytilus galloprovincialis* are a residual product of a large-scale food industry. In these ecologies the mussels form part of manifold socioecological systems. Farming is closely tied to local livelihoods and identities and sustained by intergenerational knowledge. At the same time, mussel farms also influence water quality. Through their filtering activity, they settle suspended particles from the water column as organic deposits, increasing nutrient availability and supporting microbial and invertebrate life. In this way, mussel farming decreases water turbidity, allowing greater light penetration and supporting photosynthetic life, thereby affecting sediment dynamics and ecological connectivity. By participating in the lagoon's own cycles, mussel farming forms part of its ecological balances. Local Resource / Collective Knowledge asks how architecture may similarly nest into these critical, interdependent ecosystem interactions. It does so by expanding the ideation of malleability and repair to become an integral ontological concern of the ongoing, dynamic relationship between the built environment and our surrounding world.

↙ **CITA, GXN, and 3XN, Local Resource / Collective Knowledge, 19th International Architecture Exhibition, *Intelligens. Natural. Artificial. Collective*, Venice Architecture Biennale, Venice, Italy, 2025**

Cladding panels in the installation combine up to three different blue biopolymer composite recipes, which vary the ratios of blue mussel shell and eelgrass. Panels are packed by hand, and the thickness of each material band is determined bottom up by the maker. The panels are produced in three different thicknesses and milled using different CNC patterns.

↓ The blue mussel waste that forms the basis of the blue biomass tiles is residual to the local mussel farming industries.

Transformative Potentials

The metabolistic resource thinking exemplified in these three case studies takes as its point of departure a critical examination of the prevailing conceptualization of biogenic materials in architecture. By challenging the assumption that bio-based resources are inexhaustible in their renewability, it argues that current circular design frameworks are limited in their understanding of end points and that by questioning our models of resource procurement and material systems thinking, a new transformative understanding of the biogenic can be imagined.

Rooted in a reconceptualization of the regenerative, the three projects explore renewable materials through the concepts of malleability, repair, and ecological belonging to shape an architecture that is adaptive and deeply embedded within ecological cycles. They point toward a more richly biogenic architecture in which its material systems are composed and understood through their ability to absorb regenerative and residual resource streams. By viewing resource procurement as dynamic and evolving processes, we can move toward an architecture that must exist within frameworks of regenerative cycles and engage in processes of repair and adaptation.

These case studies explore how reparatory practices can be mutually operative. On the one hand, they suggest building as an ongoing activity continuously addressed by, and adjusting to, incoming resource streams, while on the other hand pointing to architecture's direct, remediate role within its containing ecology. In this way, a novel conceptualization of extended lifespan emerges, where buildings and materials are no longer expected to function within linear, fixed timelines, but instead exist within frameworks of regenerative cycles.

1 The research has received funding from the European Research Council (ERC) under the European Union's Horizon 2020 Research and Innovation Programme (grant agreement No. 101019693), the Building with Blue Biomass global innovation network supported by the Danish Ministry of Higher Education and Science, and the foundations: Realdania, Danish Arts Foundation, Knud Højgaards Fond, and Beckett Fonden. The projects have been supported through a wide range of collaborations: Technical University of Denmark, Università Iuav di Venezia, the Computation in Architecture Masters programme at the Royal Danish Academy—Architecture, Design, Conservation, Havhøst, Vejle Fjordhave, Sustain.me, and Consorzio di Tutela della Cozza di Scardovari DOP.

2 Rhizerial was exhibited at *In the Making—Artistic Methods in Architecture and Design*, Royal Danish Academy—Architecture, Design, Conservation, Copenhagen, 2024–25. Team: P. Nicholas, M. Ramsgaard Thomsen, G. Rossi, and C. Eppinger.

3 Repair Ecologies was exhibited at the Danish Design Museum, Copenhagen, in 2025. Team: M. Ramsgaard Thomsen, P. Nicholas, R. S. Chiujdea, C. N. Preston Ichniowski, K. Toumpektsi, C. Eppinger, K. Sonne, C. Nelson, S. Akbari, and M. Tamke.

4 Local Resource / Collective Knowledge was exhibited at the 19th International Architecture Exhibition, *Intelligens. Natural. Artificial. Collective*, Venice Architecture Biennale, 2025. Team: Local Resource / Collective Knowledge is a collaboration between CITA, GXN, and 3XN. CITA team: P. Nicholas, M. Ramsgaard Thomsen, C. N. Preston Ichniowski, G. Kohlhaussen, L. Fritsch, and S. Løkke.

CHASING FUNGI
PERILS AND LESSONS FROM A MYCO-CRA

Jonathan Dessi-Olive

TSMAN

It was just after 11:00 p.m. on a cold January night when I pulled up to the Cartoon Network Hotel in Lancaster, Pennsylvania. I shut off my white Toyota 4Runner and sighed. Earlier that day I was at the University of Virginia (UVA) working with students on an installation called La Parete Fungina,[1] for the *Biomaterials Building Exposition*,[2] which brought together architect-scholars who work with renewable, carbon-sequestering biomaterials. My time working with students was limited to a five-day workshop. On the second day, we learned that due to shipping delays our 650 lb. (300 kg) of living materials were going to arrive after my departure from Charlottesville, Virginia. The pallet was stuck near Washington DC and the only way for us to obtain the materials before the end of the week was to fetch them ourselves. The supplier gave me a location 230 miles (400 km) away in Westwood, Pennsylvania, and by evening, I was driving north in my rented SUV. I had a friend book a hotel in the nearest city. They only texted me the address, so this television-themed motel in the middle of Amish country was both a surprise and a welcomed gesture to bring light to the situation. My room was a flutter of pink and blue and yellow and hyper-graphic wallpaper. The bedspread was Gumball and Darwin Watterson. My shower curtain was Powerpuff Girls. Despite the loudness of cartoon ephemera, I managed to get about six hours of sleep. Time to pick up a truck-full of fungus.

↰ **MycoMatters Lab, Omid Oliyan, Rebecca Buntrock, and students at the University of North Carolina at Charlotte (UNC Charlotte), Phoenix, Charlotte Art League, North Carolina, 2023**

Phoenix is an acoustical retro-fitting sculpture that was installed at the Charlotte Art League in 2023. The sculpture is composed of sixteen uniquely shaped sheets that hang from the ceiling of the performance venue, softening and clarifying the acoustics in the space.

← **MycoMatters Lab and students at the University of Virginia, La Parete Fungina, Charlottesville, Virginia, 2022**

La Parete Fungina shown in the snow at the inauguration of the *Biomaterial Building Exposition*, organized by Katie MacDonald and Kyle Schumann, at the University of Virginia. The wall prototype was an assemblage of living V-shaped slabs of myco-materials that were cultivated into a monolithic whole.

Mycelium Architecture is NOW

Since the term "mycotecture" was coined around 2010 by American artist Philip Ross[3] (who later founded the company MycoWorks), interest in the role mycelium composite materials (myco-materials) can play in architectural assemblages has grown exponentially. Between 2016 and 2023 global publication and patent trends grew tenfold, with myco-materials outpacing other fungal products like enzymes and antibiotics.[4] Myco-materials offer a means of decarbonizing architecture through the cultivation of fungal hyphae into cellulosic fibers. Fabricating with myco-materials (myco-fabrication) is akin to the casting arts and produces lightweight materials that resemble styrene when dried. Moreover, myco-materials have tunable performative properties, can be cultivated into innumerable forms, and are compostable if so desired.

Today, myco-materials are an international enterprise and produced by companies such as Ecovative and MycoWorks in the United States, Biohm in the United Kingdom, Mogu in Italy, and MYCL in Indonesia. Ecovative has arguably had the most pioneering impact, scaling-up its production by partnering with mushroom farming industries. The company outsources the cultivation of its spawn materials to industrial mushroom growers, especially in the neighboring state of Pennsylvania, where over 60 percent of all mushrooms in the United States are grown.[5] This capacity to cultivate reliably at high volume facilitated the production of over ten thousand bricks for Hy-Fi tower by David Benjamin (The Living) and Arup. The twelve-meter-high compostable structure was debuted at MoMA PS1, New York City, in 2014. Ecovative also sells its materials to artists, researchers, and other mycophiles through platforms such as Grow.Bio (Grown.Bio in Europe). The organization's material uses a hemp substrate inoculated with a fungus from the phylum Basidiomycota, whose fruiting bodies resemble those produced by *Ganoderma lucidum* (reishi).

Large-scale myco-projects have gone from bespoke and mythical to now almost expected at design events and biennales. Recent highly visible and publicized projects have been included in the 2024 AM Summit in Herning, Denmark (by Bjarke Ingels Group), the 2025 Milan Design Week (by Henning Larsen), and the 2025 Venice Architecture Biennale (by Anomalia). Yet amongst these diversely shaped pavilions, their descriptions essentially say the same thing: namely, that they are "demonstrating the potential" of myco-materials as alternative construction materials. Rarely is any information or visual documentation of the process shown publicly or later published. This might explain the near ubiquity of pavilions—most using myco-materials as modular units—held by wood or steel frames. In terms of building methodologies, "myco-tecture" has remained rather unchanged and has not advanced significantly beyond the realm of "potential" in the ten years since Hy-Fi, other than maybe with the shapes of the modules.

↰ **MycoMatters Lab and students at the Georgia Institute of Technology, Monolito Micelio, Atlanta, Georgia, 2018**

The Monolito Micelio is the largest example of a monolithic mycelium structure to date. It was cultivated with students in spring 2018 for a barbershop quartet performance at the Georgia Institute of Technology.

↙ **MycoMatters Lab and students at the Georgia Institute of Technology, packing the Monolito, Atlanta, Georgia, 2018**

The process of cultivating the Monolito Micelio involved packing 180 bags (800 kg) of living material sourced from Ecovative into a large wooden formwork. The structure took about five days to grow and served as a performance pavilion.

↙ **MycoMatters Lab, 4Runner full of fungus, Westwood, Pennsylvania, 2022**

The materials for La Parete Fungina were procured by collecting them from a supplier in a rented Toyota 4Runner. The 65 bags (300 kg) were packed into the SUV and driven 230 miles from Westwood, Pennsylvania, to Charlottesville, Virginia.

On the Road and Back Again

The sun was rising as I left the Cartoon Network Hotel and headed to the supplier, Lambert Spawn. While driving, I thought about how the challenges associated with procuring large quantities of live myco-materials were not new to me. In my time building with fungi, I have always sourced materials from Grow.Bio/Ecovative, for better and sometimes for worse. We had a similar shipping issue in 2018 when I was prototyping monolithic mycelium arch structures with students at the Georgia Institute of Technology. In response, an intern at Grow.Bio named Shane flew from Albany, New York, to Atlanta, Georgia, and hand-delivered the materials in four large second-hand suitcases that he also left with us. Thankfully, when we ordered 180 bags (around 800 kg) a month later, in April 2018, they were ready for us and the delivery went ahead without a hiccup. Those materials were cultivated into Monolito Micelio,[6] a vaulted structure that served as a temporary performance pavilion for a barbershop quartet at the Georgia Institute of Technology.

The structure demonstrated that non-modular assemblages of myco-materials are possible, but also that monolithic mycelium is hard work, risky, and expensive. It took the students and me over twelve hours to process the materials and pack the formwork. Despite being cultivated outdoors, it succeeded (in part) because the materials are forgiving to novice cultivators and grow quickly enough to bear resilience to contamination. The quality of the materials we used allowed us to work less like biologists and more like we were accustomed to using conventional, inert materials. All this to say, the drive to fetch these materials was worth it.

Before long, I arrived at Lambert Spawn. The part of the facility I saw was a refrigerated warehouse with endless shelves of plastic grow bags of various fungi in their preferred growing medium. Most would be sent to commercial mushroom farms and cultivated into food. The warehouse attendant located my pallet and rolled it out to my 4Runner. “In *there*?” he asked, looking a bit baffled. I nodded as I folded down the back seat. “Okay,” he chuckled nervously while cutting the shrink wrap enveloping my bags. Two at a time, we loaded 300 kg of living material into the back of the SUV. Normally, Lambert Spawn would have freight-shipped these materials next-day in refrigerated containers to slow the fungi from over-foraging and forming a hardened mass. For most of the drive back I had the windows down, with air-conditioning blasting, to keep the bags as cold as possible. Back at UVA, the students and I completed the laborious task of breaking up the biomass and repacking it into formworks over the next few days. The parts grew and La Parete Fungina was installed on time. Crisis averted.

Myco-Fabrication for (Almost) Everyone

Myco-fabrication is interesting, in part, because fungi are challenging to work with. They require a different sense of craft, care, and attention throughout each phase of their production compared to common, inert materials. Beyond the cost of the material itself, it is both energetically and financially costly to hire refrigerated transport and storage. Cultivating myco-materials can produce waste, including plastics for formworks, growing environments, and PPE. To produce parts reliably, cultivators need to keep their working and growing environments clean and often require active temperature and humidity controls. After cultivation, drying large parts can be challenging without sizable ovens or solar kilns.

Overcoming these challenges at large scale has been the underlying focus of the MycoMatters Lab. To curb up-front material costs and avoid shipping and storing pallets of living materials, we now order only as many 20 lb. bags (9 kg) from Grow.Bio as we can stock in our refrigerator—which is around twelve. In nearly all applications in the Lab, we propagate the purchased material into *unsterilized* substrates in large bins to increase our overall volumetric yield. Our recipe is:

10 lb. (4.5 kg) dry unsterilized hemp bedding
4.5 lb. (2 kg) living Grow.bio hemp spawn
8 cups (1 kg) kitchen flour
2 +/- ½ gallons water (7–8 L)
Cover the substrate with food-safe plastic film and poke holes
5–7 days growing time in temperatures of at least 70°F (21°C)

Growing large bins of materials allows for myco-fabrication in semi-clean or shared work environments and can expand or contract based on varying need. Bin propagation was important to the production of Phoenix, an acoustical retrofitting sculpture installed in a performance space at the Charlotte Art League in 2023. Instead of working with massive amounts of living materials all at once, as we had with both the Monolito Micelio and La Parete Fungina, here we cultivated eighteen bins in phased batches, using the recipe above. The idea was to relocate some work off-site and spread out the labor over several weeks. Phoenix needed 300 kg of material to produce but only used 36 kg of purchased living materials. In comparison, La Parete Fungina used 300 kg, and Monolito Micelio used 800 kg.

Phoenix was created using a technique called *hang-dry*,[7] which permits the forming of myco-materials into expressive three-dimensional curvatures otherwise wasteful or not possible when using casting or formwork-based approaches. The living materials were formed and cultivated into flat two-dimensional sheets up to six meters long. After several days, during the intermediate stage of growth when they were cohesive and gelatinous, the sheets were hung from precise support points, bent into form, and dried in place. Importantly, the technique benefits from augmented reality assistance with Fologram and a HoloLens headset, which helped translate complex, digitally generated geometries into the physical materials. While there were numerous efficiencies in terms of material cost, quantity, and waste, the process of producing Phoenix was still long and laborious.

Myco-Hybrid Fabrication is NEXT

In the last few decades, the possibilities and limits of myco-materials have been widely explored. Modular applications of myco-materials will likely continue to be appropriate for applications such as large-scale acoustical arrays, interior furnishings, and ephemeral structures such as scenography. On their own, however, there are clear limits to what myco-materials can do. To avoid the perpetual repetition of the mycelium pavilion, the next avenue of research must focus on biohybrid materials and bio-adhered layering.

In a recent experiment, students enrolled in the "How to Grow a House" design studio used a monolithic myco-material formwork for a concrete floor slab prototype. As a stay-in-place formwork system, myco-materials were imagined to be leveraged for their isolative properties, while the vaulted form minimized the use of concrete. Monolithic mycelium has its limits though, so for such a system to achieve building scale it would have to evolve to include assemblages of mycelium parts, and to ensure proper curing, there would need to be a coating between the mycelium and concrete (maybe wax).

While the projects presented above stray from common modular approaches to myco-fabrication, they are also guilty of being little more than "demonstrating the potential" of myco-materials. How myco-materials will best serve architectural assemblies, whether ephemeral or permanent, is still an open question. Mycelium can and will grow onto, into, or through most porous substrates given the right conditions. There are limitless possibilities of what to explore next, including hybrids with porous materials such as wood, ceramics, textiles, and other nonconventional or emerging biomaterials.

← **MycoMatters Lab and students at UNC Charlotte, process of making mycelium sheets, Charlotte, North Carolina, 2023**

The sixteen flowing sheets that composed Phoenix were formed and cultivated flat on-site at the Charlotte Art League from materials cultivated off-site in bins. Each sheet had a unique form and was produced by bending a reusable formwork system guided by an augmented reality headset.

← The process of hanging living mycelium sheets for Phoenix involved picking the sheet up fromw the ground, bending it into shape, and suspending it to dry. Numerous people were needed to form the sheets, which were up to six meters long.

↑ **MycoMatters Lab and students at UNC Charlotte, Myco-Concrete Floor Prototype, Charlotte, North Carolina, 2024**

A small prototype for a sustainable floor system was produced in spring 2024 by graduate architecture students attending the "How to Grow a House?" design studio at UNC Charlotte. The prototype uses a monolithic mycelium formwork for a vaulted concrete slab, demonstrating a possible trajectory in myco-hybrid fabrication techniques.

1 Jonathan Dessi-Olive, "Strategies for Growing Large-Scale Mycelium Structures," *Biomimetics* 7, no. 3 (2022): 129.
2 "Biomaterials Building Exposition," School of Architecture, University of Virginia, March 14–April 30, 2022, https://www.biomaterialbuilding.com/.
3 Kimberley Mok, "Mycotecture: Building with Mushrooms? This Inventor Says Yes," *Treehugger*, September 26, 2012, https://www.treehugger.com/mycotecture-mushroom-bricks-philip-ross-4857225.
4 Nungnit Wattanavichean, Jakkapon Phanthuwongpakdee, Preeyaporn Koedrith, et al., "Mycelium-Based Breakthroughs: Exploring Commercialization, Research, and Next-Gen Possibilities," *Circular Economy and Sustainability* (April 2025).
5 "Mushroom Production in Pennsylvania," PennState Extension, accessed August 29, 2025, https://extension.psu.edu/forage-and-food-crops/mushrooms.
6 Jonathan Dessi-Olive, "Monolithic Mycelium: Growing Vault Structures," in *Proceedings to the International Conference on Non-Conventional Materials and Technologies (NOCMAT)*, July 2019, Nairobi, Kenya.
7 Jonathan Dessi-Olive, Omid Oliyan Torghabehi, and Rebecca Buntrock, "Computational Design and Craft of Phoenix: A Hanging Myco-Structure," in *ACADIA 2023: Habits of the Anthropocene; Proceedings Two*, ed. Assia Crawford, Nancy Diniz, Richard Beckett, Jamie Vanucchi, and Marc Swackhamer (Denver: ACADIA, 2023), https://papers.cumincad.org/data/works/att/acadia23_v2_184.pdf.

Ronald Rael

Digital Earth Nurturing the Biological and **Ecological**

↰ **Ronald Rael and Virginia San Fratello (Emerging Objects), Casa Covida, La Florida, Colorado, 2019**

Casa Covida's inflatable roof, resembling a blooming cactus, offers ephemeral shelter from rare precipitation in an arid landscape that receives an average of just 228 mm of annual rainfall.

← **Ronald Rael and Virginia San Fratello (Emerging Objects), Casa Covida, La Florida, Colorado, 2019**

Beetle-kill pine, a material shaped by climate change, forms the door and lintel at Casa Covida, while a 3D-printed mold casts the handle from recycled aluminum.

→ **Ronald Rael, Terrano, La Florida, Colorado, 2024**

Terrano's interior demonstrates a stabilized microclimate, contrasting sharply with the extreme temperature fluctuations of the surrounding environment.

↘ **Ronald Rael, Adobe Oasis, Palm Springs, California, 2025**

Additive manufacturing enhances adobe's inherent benefits—low cost, energy efficiency, fire resistance, and nontoxicity—demonstrating its potential for sustainable architectural innovation.

Across continents and centuries, earthen architecture has demonstrated ecological harmony, and contemporary designers must revisit the material in the face of climate and cultural challenges.[1] Clay may be inert; it is not itself alive, but in construction it behaves almost like a living material. Unfired earthen walls breathe, erode, and eventually return to the earth.

They regulate humidity naturally, "breathing" moisture in and out to keep interiors healthy. Thick adobe and rammed earth walls act as thermal batteries, absorbing daytime heat and releasing it slowly at night to maintain stable indoor temperatures. This thermodynamic performance can eliminate or reduce the need for mechanical cooling and heating in many climates, drastically cutting energy use. In hot, arid regions especially, traditional clay houses have long stayed cool by day and warm by night through this passive regulation.

Earthen buildings also excel in fire resistance and toxin-free sustainability. Dried mud walls do not burn—an invaluable benefit as wildfires threaten many communities. Unlike petrochemical-based materials, clay is nontoxic: It contains no volatile organic compounds (VOCs), formaldehyde, or plastics to off-gas into living spaces. An adobe house is essentially made of soil and fiber, often a mixture of sand, silt, clay, water, and straw, shaped into bricks or monolithic walls. The result is a home with almost no synthetic ingredients. The largest home builders today churn out houses covered in latex paints, foams, and high-VOC materials that don't biodegrade and are often flammable. However, there is a new way—which is a very old approach. In clay construction, that old-new approach yields homes that are fundamentally healthier for inhabitants and the environment alike.

Perhaps most remarkable is the way clay structures age and participate in circular ecology. With proper care—a good roof, deep eaves, and periodic replastering—adobe buildings can last for centuries, if not millennia; countless buildings of all typologies in every climatic zone around the world stand as proof. Yet if abandoned to the elements, an earthen structure doesn't end up as indigestible rubble in a landfill. Instead, it simply melts back into the ground. Sun-dried clay will gradually disintegrate in rain and snow, returning to the soil from which it came—a sort of soil bank. In the meantime, its crumbling walls may host moss and lichen or provide nesting nooks for small creatures. When a modern concrete or steel building reaches end of life, we face a toxic demolition problem; when a clay building does, we get new soil. In this way, earthen architecture closes the loop of environmental circularity. It epitomizes cradle-to-cradle design—buildings that emerge from the earth and ultimately give back to it.

Little wonder that an estimated one-third to one-half of the world's population continue to live or work in buildings made of earth. From millennia-old earth villages in the United States to multistory cities in Yemen, clay has proven its worth as a durable, climate-responsive, and regenerative building medium. However, these long-standing building traditions are quickly being replaced by contemporary materials, such as plastics, concretes, and other synthetic and industrial products, which have proven harmful to the global environment. But recent developments in earthen construction that employ robots for material deposition are leading to innovations that reintroduce earthen architecture heritage practices into contemporary building practices. These simultaneously learn from the past while creating novel architectural possibilities for the future.

Innovations in Earthen Construction in Response to the Pandemic

One of the earliest projects heralding clay's contemporary comeback was Casa Covida in 2020.[2] Conceived during the COVID-19 pandemic, Casa Covida is a compact adobe dwelling that merges ancient and contemporary fabrication techniques. Located in the high alpine desert of southern Colorado, it was built using a robotic 3D printer to deposit coils of mud in three programmed frustrum spaces. The design was an experiment in combining 3D printing with Indigenous and traditional building materials, born of necessity when human labor and travel were limited, and it uses robotic technology created to work with the oldest material at hand: the ground itself. Casa Covida takes the form of three conjoined cylinders open to the sky. Each cylinder serves a simple function—one is a sleeping nook lined with Churro sheepskins, another a bathing space embedded into the earth that draws ancient water from a deep aquifer, and the third an open-air hearth for cooking and warmth.

↰ **Ronald Rael, Terrano, La Florida, Colorado, 2024**

As the first in situ 3D-printed adobe roof, Terrano merges ancient earthen construction with advanced fabrication, redefining structural possibilities for earth-based architecture.

→ **Ronald Rael, Terrano, La Florida, Colorado, 2024**

A woven deposition pattern ensures structural cohesion between layers, mimicking natural growth techniques while enabling complex geometries.

By stripping architecture down to elemental experiences (earth underfoot, sky overhead, fire at the center), the project harkens back to humanity's earliest clay dwellings, even as it deploys cutting-edge automation. The use of 3D printing vividly recorded the digital construction process, encoding the construction itself but also recording the weather, much like the growth rings of a tree, since rainy days allowed fewer layers, while warm and windy days had more. Casa Covida is the first prototype of a multi-chambered raw-earth house ever constructed. The project's very name, blending *casa* (home) with COVID, hints at its dual identity. It is both a shelter in a time of global crisis—not only the health crisis of the pandemic but the larger environmental crisis of the time—and an experiment in resilient cohabitation (co-*vida*) through new technology. Built in a moment when a global biological crisis had driven people apart, this little adobe retreat suggested a way to reconnect architecture with the healing rhythms of nature—to cohabit with the land, safely and sustainably.

Roofed Adobe for Remote Living

Another project, Terrano,[3] set its sights on a long-standing challenge in both earthen construction and 3D printing: the roof. Modern architects have struggled to create fully 3D-printed homes because spanning a roof without formwork is difficult.

Terrano tackled this head-on. Completed in 2023, it is perhaps the first in situ 3D-printed roofed structure ever completed, using humankind's most humble material, raw earth. The key was reviving the Nubian vault method—a millennia-old approach that builds a self-supporting earthen vault without any wooden support. The robot was programmed to deposit courses of adobe in a parabolic arch, building from an apse and ultimately resulting in a vaulted funicular structure entirely constructed of continuous coils of mud. Terrano demonstrates the capacity to move into remote environments to build walls and roof using only the ground itself, as a low-impact method of construction. This has game-changing implications and points to a future where architecture could be achieved with zero imported materials, minimal waste, and minimal skilled labor—truly building with the land and of the land.

Technically, printing a roof of raw earth was a remarkable first. And culturally, Terrano is just as significant. It closes the conceptual gap between high-tech building and vernacular wisdom. The Nubian vault, invented by Indigenous builders centuries ago, was essentially taught to a robot, and in doing so, Terrano bridges a continuum from ancient mud masonry to autonomous construction. The structure's corrugated texture prevents erosion from rain and snow as well as creating a stronger material bond between layers. As an enclosed shelter, it offers slightly more protection from the weather than its predecessors, providing a preview of how future clay homes could fully enclose space while remaining breathable, bio-responsive, and beautiful.

Land Art Meets Climate Action

Adobe Oasis[4] carries a strong message: Earthen architecture is a sophisticated, sustainable solution for the climate crisis. Its design bridges the Indigenous building traditions of desert oases with advanced robotic fabrication, embodying the ethos of biodesign as a climate-responsive endeavor. Given it is low cost, energy efficient, fire resistant, and nontoxic, digital earthen construction has significant advantages—an important consideration in the aftermath of the Los Angeles fires, which left large populations of the city without homes, and a toxic soup of burnt building materials in the fire's wake. By leveraging robotic technology, these benefits are enhanced by advancements in additive manufacturing, paving the way for sustainable architecture. Adobe Oasis was powered by solar energy and demonstrated innovation in architecture operating on the elemental—earth, wind, sun, and water.

The project engaged with deep time and heritage, connecting ten thousand years of earthen building history while offering a counterpoint to the environmental harm of contemporary construction. Walking through its earthen corridors, one can feel that connection—the thickness of the walls muffling sound, the earth's coolness radiating comfort in the 110-degree-Fahrenheit heat. And as a temporary structure, the goal is to recycle Adobe Oasis to produce new structures. In a region plagued by fire, drought, and development, Adobe Oasis appeared as a powerful alternative, highlighting the potential of buildings to be constructed from dust and return to dust as a template for rebuilding Los Angeles—where dust has instead, in areas of fire devastation, become a biohazard.

↑ **Ronald Rael, Adobe Oasis, Palm Springs, California, 2025**

The aerial perspective shows the structure harmoniously embedded in the desert, its form responding to both environmental conditions and material possibilities.

→ **Ronald Rael, Skylos, La Florida, Colorado, 2022**

This aerial view reveals the eight sky-viewing chambers of Skylos, a structure designed to frame celestial perspectives while integrating with the high desert landscape.

→ The oculus frames a dynamic aperture to the sky, emphasizing the structure's role as a mediator between earth and atmosphere.

Earthen Cycles Under the Stars

Completed in 2022, Skylos[5] is a permanent installation of eight conjoined adobe silos arranged in a ring. It was the first 3D-printed earthen building in the world to be permitted for construction as well as the largest such structure to date. This was a watershed moment: clay architecture, digitally crafted, proving its viability to regulators and the public.

Named after the abundance of silos in the region (and a play on "sky" and "silos"), Skylos holds not grain but the sky, and takes the form of eight hefty cylinders made from locally harvested mud and straw. Each cylinder forms a collection of observation chimneys for stargazing at eight thousand feet elevation. Visitors step inside a cool clay chamber and gaze up at a perfectly round patch of firmament. In an era of sealed, air-conditioned boxes, Skylos instead prioritizes elemental experience—earth below, sky above, people gathering in between.

Thick earthen walls are printed in horizontal bands, with a pattern almost like basket-weave where the mud extrusions overlap. Pieces of straw poke out here and there, catching the sunlight and reminding you that these walls are made partially of organic matter—straw, like clay, is part of the natural life cycle. Through the cold winters and hot summers, its earthen mass buffering temperature swings. It also serves as a way of reinstituting local knowledge and traditions to assure its longevity—future earthen plaster maintenance, for example, could spark a revival of mud plastering skills among local communities. Establishing a reconnection to heritage through forward-thinking design, Skylos closes the loop between past and future, nature and technology.

Scaling Clay for a Sustainable Future

The through-line connecting Casa Covida, Terrano, Adobe Oasis, and Skylos is that each was realized using the technology platform Muddy Robots,[6] which combines ancestral intelligence, additive manufacturing, and artificial intelligence (AI[3]). Portable 3D printers powered by solar energy, and loaded with site-specific clay, allow for buildings that are low emission, rapid to assemble, and embedded in local ecologies.[7]

Muddy Robots is not speculative—it is operational. The system enables rapid, sustainable construction for housing, cultural spaces, and disaster relief. It avoids toxic chemicals, minimizes embodied carbon, and challenges global construction norms.

While the promise of perpetual architecture may still be evolving, what's already clear is this: Earthen architecture as digital earth is not about nostalgia. It's about urgency. It connects past to present, land to structure, and code to craft. By weaving clay's living qualities with technological tools, it offers a blueprint for architecture that not only survives the climate crisis but helps heal it.

1 Ronald Rael, *Earth Architecture* (New York: Princeton Architectural Press, 2008); and Ronald Rael (founder), *Earth Architecture: Architecture, Art, Design, and Culture using mud, clay, soil, and dust* (blog), https://www.eartharchitecture.org.
2 Casa Covida by Ronald Rael and Virginia San Fratello (Emerging Objects), 2019. Project team: Ronald Rael, Virginia San Fratello, Mattias Rael, Sandy Curth, and Logman Arja.
3 Terrano by Ronald Rael, 2024. Project team: Ronald Rael, Barrak Darweesh, and Mattias Rael.
4 Adobe Oasis by Ronald Rael, 2025. Project team: Ronald Rael, Andrew Kudless, Hector Velderraint, and Jamie Miranda; and Ronald Rael, "Desert X 2023: Adobe Oasis" (exhibition statement), Desert X 2023 (Palm Springs, California), 2023.
5 Skylos by Ronald Rael, 2022. Project team: Ronald Rael, Logman Arja, Michael Jiron, Elliot Surber, and Mattias Rael; and "Skylos" (project description), Emerging Objects, https://www.emergingobjects.com.
6 Ronald Rael, "About—Muddy Robots," Muddy Robots, https://www.muddyrobots.com.
7 Jaxson Stone, "Ronald Rael's Solution to the Housing Crisis? Muddy Robots," *Metropolis Magazine*, April 9, 2025.

A6

Forming a New Ruskinian Relationship

An architect should live as little in cities as a painter. Send him to our hills, and let him study there what nature understands by a buttress, and what by a dome.

—John Ruskin[1]

Neil Spiller

The Architectural Association (AA) has a rural outlier, a 150-hectare (370-acre) forest park campus in the English county of Dorset. Contrary to the studied urban and Georgian sophistication of its London headquarters in Bedford Square, Bloomsbury, Hooke Park is a place of workshops, robot arms, full-size fabrications, and student-built teaching and living accommodation, and home of the AA's Design + Make master's courses. The pedagogy here is to foster new relationships between makers/designers and material (mostly timber that has been sustainably husbanded), aided and abetted by new digital technologies. It is very much a hands-on learning-by-doing environment, asking students to work at 1:1 scale and build interventions within its woodland curtilage, thus breaking the familiar, distanced position of many architects from the actual construction of their designs.

↰ **Architectural Association, Design + Make, robotic bandsaw generation 3, Hooke Park, Beaminster, Dorset, UK, 2017**

At Hooke Park, unique production methods of fabrication often call for the necessity to make workshop tools on-site.

→ **Architectural Association, Design + Make, Woodland Cabin, Hooke Park, Beaminster, Dorset, UK, 2019**

Robotic fabrication was used to establish multiple datums along a complex form, allowing the organic geometry to act as the primary locator for the envelope. Woodland Cabin is here seen near completion, carefully nested within the fabric of the forest.

Historic Hooke **Park**

The site and its surroundings have a long history. It has evidence of Iron Age settlements and hillforts. It's been used for recreational hunting. After World War II, the site was clear-felled to help feed the hunger for building materials needed for reconstruction, and then in the early 1950s, now under the ownership of the Forestry Commission, it was replanted with a mixture of beech and Norway spruce, and some oak. Remnants of the old woods—represented by fragments of alder, willow, and hazel—were left untouched. Beech was an important commercial species feeding the large volume UK furniture manufacturers of the time, while Norway spruce is the leading structural and general use timber throughout Europe.

British furniture designer and maker John Makepeace recognized the potential of Hooke Park as an extension of his furniture design school, Parnham College, and from 1982 until 2000 Hooke Park was owned by the Parnham Trust, with the AA acquiring it in 2001. Nowadays, the site produces timber for its own use, sells timber to local sawmills, heats itself with its own timber, and preserves itself as an ecological resource for the wider Dorset landscape. The campus exhibits nearly forty years of timber experimental buildings beginning under the Parnham Trust's ownership, and subsequently the AA's, and such construction continues today. An example is the Woodland Cabin (2019). A small timber structure dealing with the challenges of uniting organic and milled geometries, demountability and reversible joinery, and the use of inherent timber properties, the cabin aims to resolve ambitious design and fabrication techniques at a domestic scale. Overall, the campus offers an environment uniquely tailored to the development of a critical, contemporary, rural-focused discourse angled toward challenges caused by the climate crisis. Studying at Hooke Park is a communal experience—living and working on-site, dining together at 1:00 PM—which helps create a convivial, inclusive, nature-focused pedagogic practice.

Biology and botany seldom, if ever, create forms in straight lines or orthogonal joints and branchings. Yet we as architects are obsessed with imposing linear arrangements on our buildings and building materials. This is a twentieth-century affectation echoing industrial conveyor belt uniformity and mass production. However, the twenty-first century has given us, through digital fabrication and dexterous scanning techniques, the ability to create bespoke architectural articulations and proposals that utilize the nonuniformity of natural forms. It is this potential that Hooke Park exploits in its creative endeavors—recognizing natural formal variation as a key component and driving imperative in its actions and outputs.

This approach resonates with some of the writings of nineteenth-century lecturer, art historian, and art critic John Ruskin, who wanted artisans to be liberated by expressing their thoughtful joy in making their work, and not thought of as mindless operatives engaged in acting out architects' instructions. He expressed this desire thus: "We want one man to be always thinking, and another to be always working, and we call one a gentleman, and the other an operative; whereas the workman ought often to be thinking, and the thinker often to be working, and both should be gentlemen, in the best sense."[2] This had ramifications for the structure and happiness of Victorian society, Ruskin thought. "As it is, we make both ungentle, the one envying, the other despising, his brother; and the mass of society is made up of morbid thinkers and miserable workers. Now it is only by labour that thought can be made healthy, and only by thought that labour can be made happy, and the two cannot be separated with impunity."[3]

At Hooke Park, facilitated by Programme Directors Emmanuel Vercruysse and Kate Davies, there is an air of neo-Ruskinian praxis supported by numerous combinations of craftmanship expertise and digital algorithmics—a new healthy analysis of beauty.

As Caitlin Duffy has observed, "For Ruskin, the classical style of architecture and the [Victorian] current industrial method both lacked evidence of individuality, unlike the gothic, which occurred in between these two periods of architecture and allowed builders to represent their imperfections and 'wolfish life' through that architecture."[4]

The constructions made at Hooke Park certainly have a quality of "wolfish life" about them—none more so than Tree and Truss (2023), a project that investigates the structural system and composition of the third phase of the ongoing student-built campus hub, Wakeford Hall, which contains a library and other student-facing functions. It was conceived to start a conversation about hybrid structural elements that seamlessly make the transition from natural geometry to engineered timber, and the design was intended to prioritize material efficiency. The structural system consists of four key components: a cedar foundation, the tree column, glulam connection, and the truss. The cedar foundation was produced through a robotic chainsaw application and the geometry evolved through in-depth studies of diverse Japanese scarf joinery. For the tree column element, research focused on combining the robotic chainsaw application with manual finishing strategies. This method strived for maximum flexibility and tolerance as it accounted for the many complexities involved with assembly and the difference in behavior between natural green timber (the tree) and engineered glued laminated timber (glulam). The glulam components required the development of a detailed knowledge of the structural requirements so the timber could be correctly configured to align with the forces it had to withstand. This also necessitated controlling the moisture content through the use of a kiln and complex timber glueing strategies, resulting in the production of specific and bespoke central nodes—produced using advanced CNC (Computer Numerical Control) operations—which connected the unique tree geometries with the trusses. The trusses themselves are composite structures using small-diameter branches and engineered timber. CNC processes were also developed to produce the members of the truss. The extreme tension load in a single location in the truss led to the development of a highly engineered timber composite, as it required the inlay of a hardwood component to reinforce the softwood.

Wolfish **Life**

→ **Architectural Association, Design + Make, Tree and Truss, Royal Academy Summer Exhibition, London, 2023**

Tree and Truss is a composite structure utilizing small-diameter branches and engineered timber. The sophistication of the built work is testament to the iterative cycles of prototyping, fabrication, analysis, and remaking as these propelled the project forward, leading to a final resolution of ambitious complexity.

Forest **Datum**

↑ **Architectural Association, Design + Make, Forest Datum, Hooke Park, Beaminster, Dorset, UK, 2024**

The production of an experimental lightweight timber structure from small-diameter (20–50 mm) forking beech branches. Forest Datum forms a jetty suspended within its woodland setting.

↘ The underbelly of Forest Datum, revealing the carefully balanced but quirky structural logic that relied on the firmness and integrity of the beech timber bifurcations from the young canopy branches.

↘ Close-up of the small-diameter crown timber tensioned within the structure—the absence of mechanical fixings allows for a rapid assembly and disassembly.

Another project that exhibits the quality of "wolfish life" is Forest Datum (2024). This uses crown timber in an experimental structural system to create a jetty in the landscape of Hooke Park, orientated along an imaginary line connecting the woodland campus to the Jurassic coast 12.5 km away. Crown timber was sourced from pruning the smaller branches high up in the forest canopy, a regular forestry management practice that improves the health of the tree and the quality of the timber yield, or as a byproduct of timber harvesting. Small-diameter beech crown timber is abundantly available at Hooke Park but difficult to use other than for chip or fuel and is considered close to zero value within the timber supply chain. The project utilizes small timber forks, a material that is inherently strong through its geometry—every branch is unique. The complexity of this unique geometry is navigated through a multi-pass cutting jig, a simple low-tech object that contains the intelligence of the structural cell parameters and defines a sequence of cutting operations that require only a bandsaw. This locates the branch within a fixed cell geometry situating each unique geometrical piece within a repeatable cell structure.

The structure is assembled and tensioned by Dyneema cable to work in compression without permanent or mechanical fixings (Dyneema cable is known for its very high strength to weight ratio). This assembly logic allows for rapid assembly and disassembly.

1 John Ruskin, *The Seven Lamps of Architecture*, new edition (London: Dover Publications, 2000), 122.
2 John Ruskin, "The Nature of Gothic," in *The Stones of Venice*, vol. 2, *The Complete Works of John Ruskin in 39 volumes*, ed. Edward Tyas Cook and Alexander Wedderburn (London: George Allen and Unwin, 1903–12). The chapter "The Nature of Gothic" appears in volume 10, p. 201, of this collection.
3 Ruskin, "The Nature of Gothic."
4 Caitlin Duffy, "The Stones of Venice (1851–53): 'The Nature of the Gothic,'" *Caitlin Duffy*, January 20, 2019, https://caitlinduffy.hcommons.org.

Natural **Respect**

Design + Make's explorations and constructions demonstrate a respect for nature, its conservation and inspiration. The combining of biological, mechanical, and computational systems with natural material geometries and textures has so much potential. The contrast between the rough-hewn and the elegantly smooth can be one of the most pleasing experiences buildings have to offer. The creation of such riotous but pure buildings and their constituent parts is a major factor in Hooke Park's unique pedagogic ambience.

WE THE BACTERIA

Beatriz Colomina and Mark Wigley

NOTES TOWARD BIOTIC ARCHITECTURE

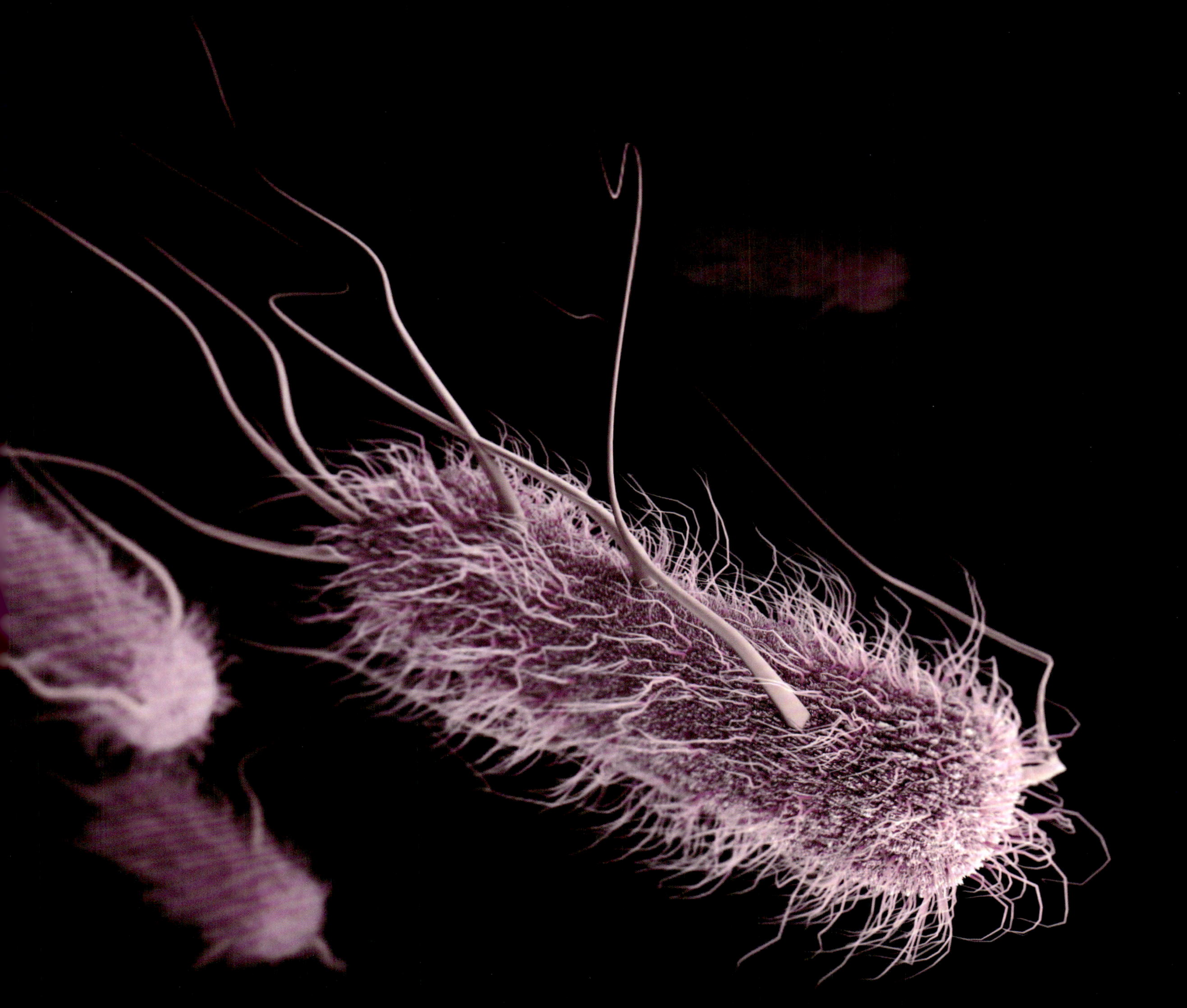

Biodesign challenges us to move beyond using biology as a mere resource and instead to see it as a partner and a model. *We the Bacteria*[1] provides a radical and urgent manifesto for this transition. It argues that the core principles of modern architecture—sterilization, isolation, and human-centric control—are fundamentally at odds with human and planetary health. By framing bacteria as the foundational lifeforms that we have always been entangled with, rather than as our enemy, it demands a new paradigm: an architecture that learns from the microbiome, embraces porosity, and promotes microbial diversity. This is biodesign at its most profound—not just using living things, but rethinking design from the ground up based on biological creativity and intelligence.

The question of health and the question of inequality have always been entangled. Inequality is about health in its broadest sense—well-being. Paradigms of health evolve, and each paradigm implies a different politics. Today, health is increasingly thought in terms of the microbiome, the trillions of microbes that inhabit us, feed us, maintain us, and defend us. Many of the diseases of our time—obesity, diabetes, cancers, autoimmune disorders, allergies, autism, Alzheimer's—are now understood as the consequence of the diminishing diversity of bacteria, and architecture is implicated. These diseases of the built environment are inseparable from planetary health—the whole biosphere.

In using architecture as a protective medical bubble, humans isolate themselves from almost all other species, as if they were a biological and existential threat. Architecture creates a division between internal and external life, a biopolitical division between different forms of life. It divides the human from the kaleidoscopic, ever-shifting entanglements of diverse species of animal, plant, and insect life, not to mention most other humans.

The first form of life was microbes, 4.2 billion years ago. They produced and sustained the whole biosphere, making plants possible, animals possible, and eventually humans. Bacteria sustain life—they are life—yet architecture treats them as an invisible enemy that needs to be exterminated.

We the Bacteria tries to think what happens when bacteria are at the center of design. This means decentering the human. Human-centered-design sounds good, but it has been terrible for humans, other species, and the planet. Besides, the human in "human-centered design" is typically white, male, Western, and solitary. Centering architecture on bacteria would actually take better care of the human but it would not be the privileged figure that sees itself as superior to all other species and most other humans.

The human is not one thing but an endlessly complex, ever-changing trans-species collaboration. The human is just a bag of bacteria, around thirty-nine trillion of them—plus at least as many viruses and other single-celled organisms—mainly in the digestive tract. Many of these microbes have been around for millions of years, while the human is a very recent arrival and might already be on the way out. The human is wrapped around the seven meters or so of the intestine of an adult. The real limit of the human is not the outside skin but the inner, moist, porous lining of the gut through which the world enters us.

Architecture has been on the front lines of the battle against bacteria, practically from the moment Louis Pasteur and Robert Koch demonstrated the link between specific bacteria and diseases like anthrax, tuberculosis, and cholera in the late 1870s and early 1880s. Sanitary reform had already dramatically transformed cities and buildings with sunlight, ventilation, plumbing, and

↰ **Alissa Eckert (medical illustrator), Centers for Disease Control and Prevention, artistic recreation of a group of extended-spectrum ß-lactamase-producing E-coli**

A 3D computer-generated image of a group of extended-spectrum ß-lactamase-producing (ESBLs) Enterobacteriaceae bacteria, in this instance, E-coli, based on scanning electron microscopic (SEM) imagery.

→ **Beatriz Colomina and Mark Wigley, *We the Bacteria: Notes Toward Biotic Architecture*, installation view, 24th Triennale Milano International Exhibition, Milan, Italy, 2025**

Entrance to *We the Bacteria* exhibition at the Milan Triennale, showing the thirty-two-square-meter size of human gut if unfolded. Exhibition design by GRACE.

→ Antimicrobial systems of Triennale building, 4.5-billion-year timeline, bacterial cloud curtain, and domestic arsenal of antimicrobial products in *We the Bacteria* exhibition.

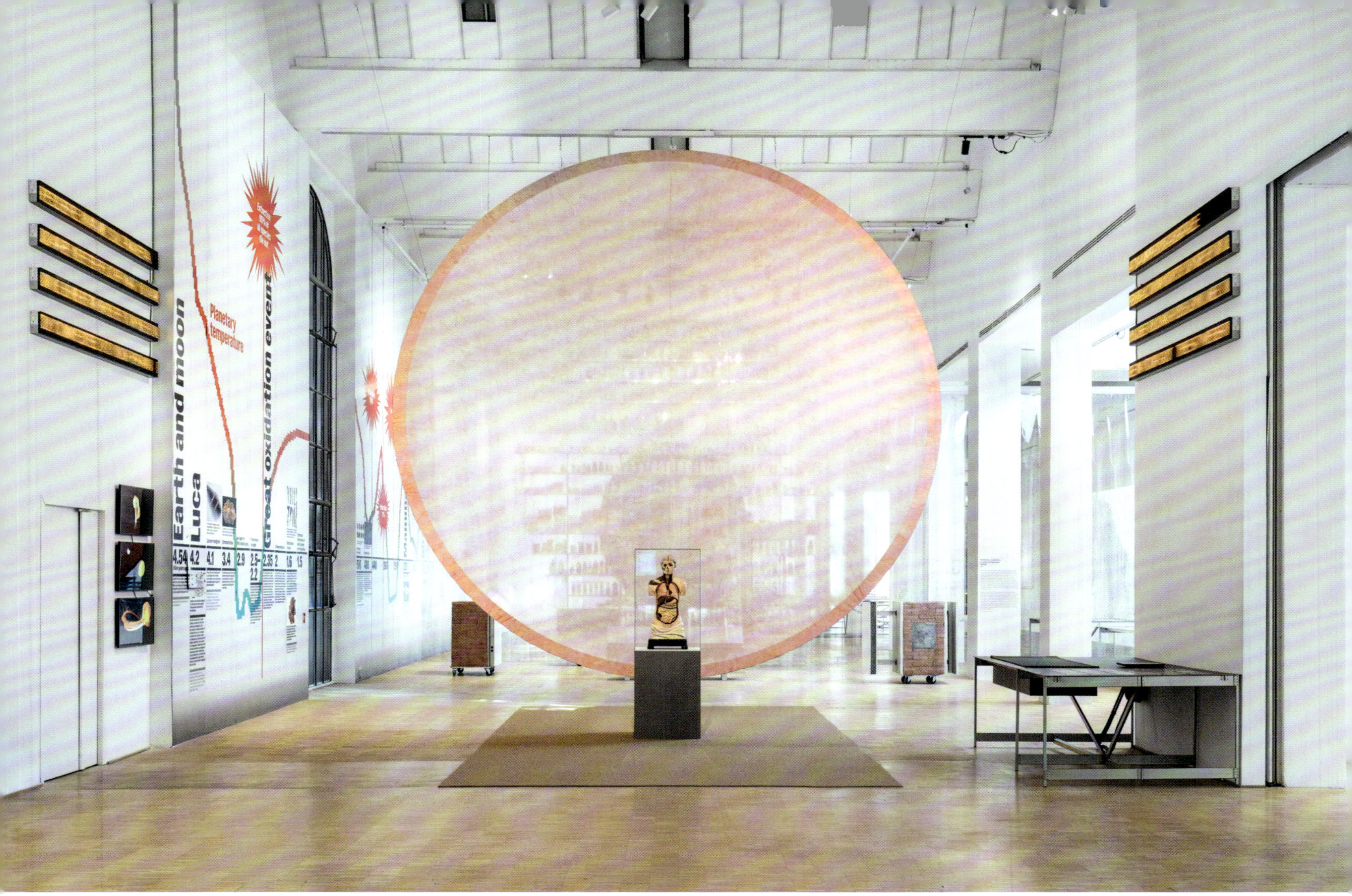
Earth and moon
Luca
Planetary temperature
Great oxidation event
4.54
4.2
4.1
3.4
2.9
2.5–2.2
2.35
2
1.6
1.5

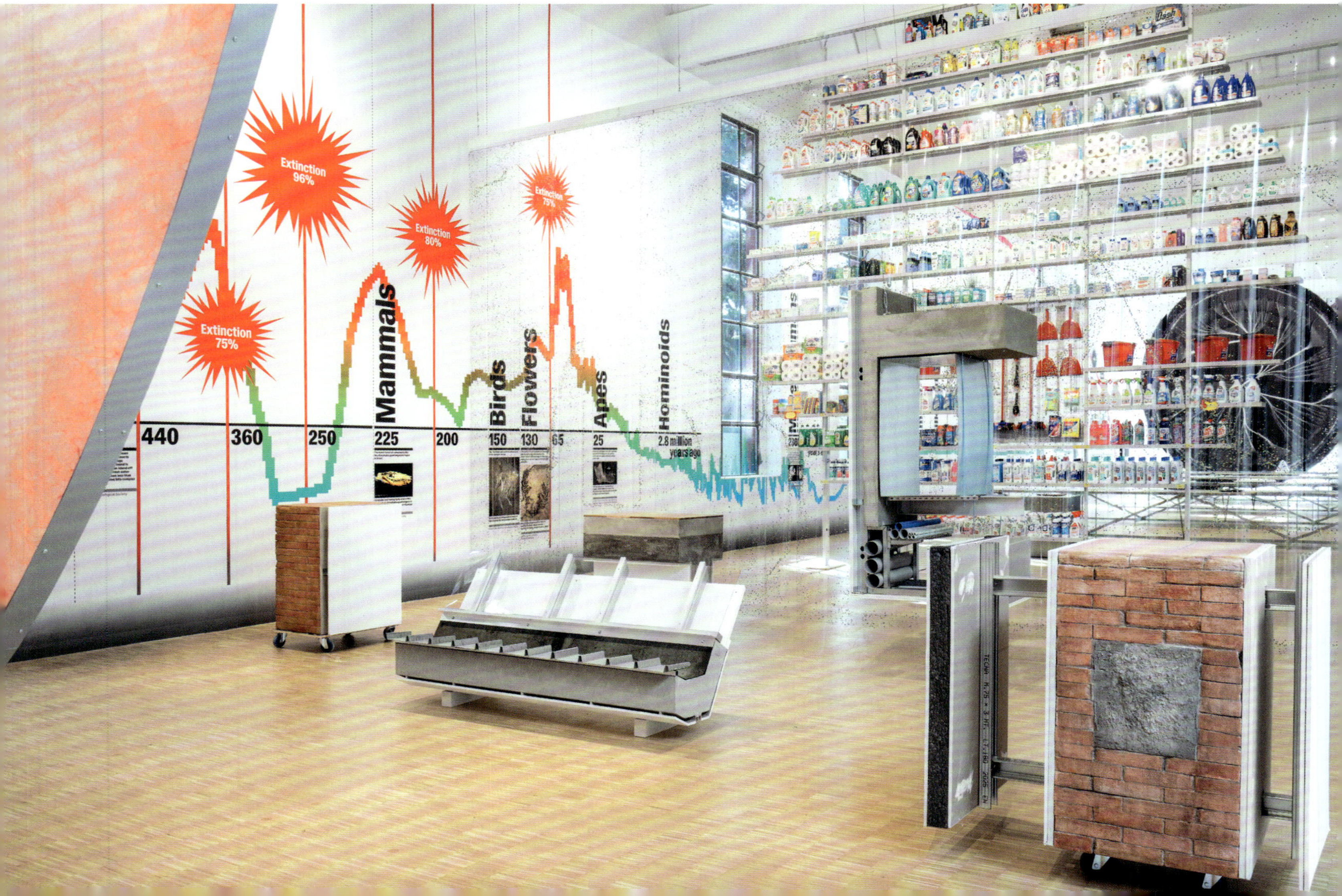
Extinction 96%
Extinction 80%
Extinction 75%
Extinction 75%
Mammals
Birds
Flowers
Apes
Hominoids
440
360
250
225
200
150
130
65
25
2.8 million years ago

hygiene throughout the nineteenth century. The campaign was now targeted on bacteria, and architecture became the first line of defense against the invisible enemy.

It was not until 1928 that the first modern antibiotic, penicillin, was accidentally discovered by Alexander Fleming, but there was already an antibiotic attitude toward buildings. In 1904, women were advised to leave petri dishes around the house to see if any bacteria had survived their cleaning routines.[2] The housewife became a bacteriologist, and the house was her laboratory. Modern architecture channeled this antibiotic approach. Le Corbusier obsessed about bacteria. He quoted counts of bacteria per cubic meter in the air, describing himself as a bacteriologist and architecture and city planning as laboratory work.[3]

Microbes were both the literal and metaphorical basis of a new architecture and urbanism. To produce the idea of modern architecture as healthy, nineteenth-century architecture was demonized as nervous, unhealthy, and literally filled with disease, especially the bacilli of tuberculosis. Modernizing architecture was firstly a form of disinfection, a purification of buildings leading to a health-giving environment of light, air, cleanliness, and smooth white surfaces without cracks or crevices where contagion might lurk. Modern architecture is only modern as long as it is free of bacteria.

Yet antibacterial architecture produces sickness. With sick building syndrome, for example, the air-conditioning systems that Le Corbusier celebrated for isolating the inside air from the contaminated outside air turned out to be reservoirs and vectors of disease. The kind of architecture that was supposed to inoculate its occupants against disease turned against them as in a science fiction movie. Extreme cleaning encourages the emergence of extremophiles. The sterilized spaces of hospitals are now plagued with bacteria resistant to antibiotics. We have even managed to send pathogens into outer space in supposedly sterilized spacecrafts.

Every architecture has produced sickness since the Neolithic move into permanently occupied buildings around ten thousand years ago. Infectious bacterial diseases like tuberculosis distorted the skeletons of the inhabitants, as has been found in recent excavations. In fact, many infectious diseases can now be traced in their DNA. Human health dramatically declined. Settled humans became shorter and lived shorter lives.

Buildings have their own microbiome and the diversity of bacteria in spaces is just as important to health as diversity in the gut. The bacteria of buildings continuously enter the human, and the bacteria of humans are spread out across buildings—along with that of fellow humans, other animals, insects, and plants. There is no strict line between body and building. In fact, a building is really an expansion of the gut. It is a way of sharing microbes with other humans and species. The microbial ecologist Maria Gloria Dominguez-Bello has shown that babies born by caesarean section are colonized by the microbiome of the delivery room rather than the mother, permanently impacting the formation of their personal microbiome, with a higher risk of immune and metabolic disorders.

Contemporary urbanized and westernized communities have only half the microbial diversity of their hunter-gatherer ancestors. Lack of diversity is a killer. The epidemiologist Martin Blaser talks about "the silent extinction of microbes," arguing that this crisis is a bigger threat to the species than climate change.[4] Antibiotic architecture both reduces microbial diversity and incubates ever more lethal antimicrobial-resistant microorganisms. The

→ **Beatriz Colomina and Mark Wigley, *We the Bacteria: Notes Toward Biotic Architecture*, installation view, 24th Triennale Milano International Exhibition, Milan, Italy, 2025**

Ten-thousand-year history of architecture tables, map of all species in human microbiome, gallery of bacteria, and scrapbook of biotic artists and architects in *We the Bacteria* exhibition.

↘ Bacterial Garden section of *We the Bacteria* exhibition, showing diverse ways in which microbes are already crucial partners in the building industry.

Flu pandemic
Polio
1918
1928
1952
1961
1891
1901
1922
1895
1928
1943
1958
1965
1981
1967
1988
1998
5000 LINES 5000 SPECIES

10.000 A
DI STOR
DELL'ARC
TETTURA

built environment weakens the ability to resist the pathogens it incubates. Deaths associated with antimicrobial resistance are predicted to rise 70 percent in the next twenty-five years to become the single largest global killer.

What would a more biotic architecture be? Probably more porous like our gut, versus the prophylactic attitude of most buildings. The immune system doesn't simply keep foreign organisms out. It curates a dynamic cosmopolitan balance. Architecture, too, can be a way of bringing others in—curating complex communities rather than narrowing them. The very idea of shelter changes from excluding danger to including it. To really shelter a species means exposing it to multiple others. The strongest defense against pathogens is other microbes.

We the Bacteria is an invitation to biodiversify spaces, techniques, companions, and attitudes. Buildings all over the world are increasingly made the same way from the same materials extracted from distant continents. This architectural monoculture needs to give way to multiple open-ended collaborations with microbes—growing building materials, maintaining buildings, conserving them, bioremediating building sites, and incubating new ways of living with others, new concepts, and new politics.

→ **Beatriz Colomina and Mark Wigley, *We the Bacteria: Notes Toward Biotic Architecture*, installation view, 24th Triennale Milano International Exhibition, Milan, Italy, 2025**

3D Petri Dish section of *We the Bacteria* exhibition, showing the living experiments of nine different studios collaborating with microbiologists.

↘ **Cover of *We the Bacteria: Notes Toward Biotic Architecture* (Zurich: Lars Müller, 2025)**

This book is a sequel to the authors' *Are We Human?* and an urgent manifesto for an alternative architectural philosophy. It treats bacteria as the real architects, construction workers, maintenance crews, and inhabitants of buildings.

1 Beatriz Colomina and Mark Wigley, *We the Bacteria: Notes Toward Biotic Architecture* (Zurich: Lars Müller, 2025).
2 Sophronia Maria Elliott, *Household Bacteriology* (Chicago: American School of Household Economics, 1904). The book was one of the volumes of the highly influential set on home economics and parts of it were serialized in *House Beautiful*.
3 Le Corbusier, *Precisions sur un etat present de l'architecture et de l'urbanisme* (Paris: Les Editions G. Cres, 1930), 144. Translation, *Precisions on the Present State of Architecture and City Planning* (Cambridge: MIT Press, 1991), 144.
4 Martin J. Blaser, "The Theory of Disappearing Microbiota and the Epidemics of Chronic Diseases," *Nature Reviews Immunology* 17, no. 8 (2017): 461–63.

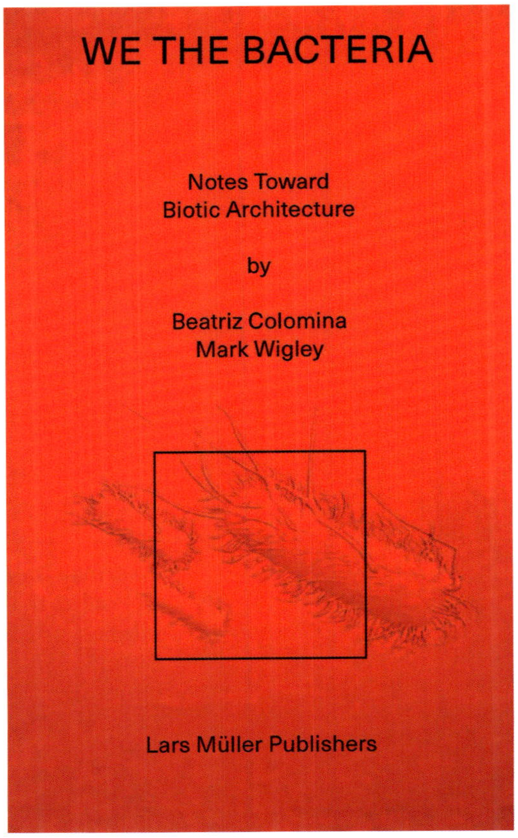
WE THE BACTERIA
Notes Toward
Biotic Architecture
by
Beatriz Colomina
Mark Wigley
Lars Müller Publishers

Rachel Armstrong is ZAP Professor of Design-Driven Construction at KU Leuven, where she leads the RAAD research team in regenerative architecture. Her work integrates microbial technologies and biomaterials to create smart biohybrid systems for sustainable construction. As lead investigator of the Microbial Hydroponics project, she explores microbial symbiosis in circular building systems. A Senior TED Fellow, Armstrong authored *Soft Living Architecture and Experimental Architecture*, which propose bio-informed frameworks for next-generation sustainable design. Her interdisciplinary approach redefines architecture as a dynamic, ecological agent, co-evolving with planetary systems to enable resilience and sustainability.

Phil Ayres is a weaver, amateur fermenter, and architect. He has held the Chair for Biohybrid Architecture at the Royal Danish Academy since it was established in 2022. The Chair investigates the applied use of weaving and fermentation for creating novel structures and materials within a built environment context. Ayres has led EU-funded research projects and collaborated extensively across a wide range of scientific, artistic, and engineering disciplines. Collaborations have also included internationally leading architectural practices such as Cave_bureau and internationally recognized cultural institutions including Louisiana Museum of Modern Art in Denmark, the Humboldt Forum in Germany, and the British Council.

Richard Beckett is an associate professor in architecture at the Bartlett, UCL, where he is director of RC7 on the B-Pro Architectural Design master's programme and leads Studio 3 on the landscape architecture MA/MLA. His research is focused on design operating at the intersection of computation, biofabrication, and microbial ecologies in buildings and cities. His research on probiotic design won the RIBA Presidents Research Award in 2021. He has built numerous projects and exhibited internationally, including at *Archilab: Naturalising Architecture*, the Pompidou Centre, and *Nature—Cooper Hewitt Design Triennial*. He is currently principal investigator on a three-year EPSRC-funded research project, "Probiotic Interventions to Reduce the Emergence and Persistence of Pathogens in Built Environments."

Beatriz Colomina is the Howard Crosby Butler Professor of the History of Architecture and the founding director of the Media and Modernity program at Princeton University. She has written and lectured extensively on questions of architecture, art, sexuality, and media. Her publications include *Sexuality and Space* (Princeton Architectural Press, 1992), *Privacy and Publicity: Modern Architecture as Mass Media* (MIT Press, 1994), *Domesticity at War* (MIT Press, 2007), *Radical Pedagogies* (MIT Press, 2022), *Sick Architecture* (MIT Press, 2025), and *Are We Human: Notes on an Archaeology of Design* (Lars Müller, 2016), with Mark Wigley. In 2025, Colomina and Wigley curated the exhibition *We the Bacteria: Notes Toward Biotic Architecture* for the 2025 Milan Triennale, a follow-up to their celebrated Are We Human? at the 3rd Istanbul Design Biennial in 2016.

Marcos Cruz is an architect and professor of Innovative Environments at the Bartlett School of Architecture, UCL, where he was director (2010–14) and established, with Dr. Brenda Parker, the Bio-Integrated Design Lab. His research and teaching (UCL, IAAC, and previously the University of Westminster and UCLA) on new forms of bio-integrated design for the built environment has focused on bioreceptive materials, poikilohydric design, and neoplasmatic architecture. He received the RIBA President's Award for Research in 2008. Cruz is co-founder of Studio Biocene and was previously co-principal of the award-winning practice marcosandmarjan (2000–15), having extensively lectured, exhibited, and published several books as well as over fifty articles.

Jonathan Dessi-Olive is an associate professor in the School of Architecture at the University of North Carolina at Charlotte (UNC Charlotte) where he teaches architecture design studios, structures, and biomaterial research seminars. He previously taught at Kansas State University, the Georgia Institute of Technology, and MIT. Dessi-Olive is also the director of MycoMatters Lab, a design, research, and education laboratory that develops novel applications of fungi-based materials in structural design, architectural acoustics, and sculpture.

Nancy Diniz is co-founder and partner of bioMATTERS, and course leader of the biodesign MA at Central Saint Martins, UAL. She has published over forty peer-reviewed publications and exhibited her work widely. She is the recipient of numerous grants, fellowships, and residencies, including the Architectural League New York Independent Project Grant 2016 and 2024, Haystack Mountain School of Crafts, Museum of Architecture and Design in Slovenia, Lisbon Architecture Triennale, Venice Architecture Biennale, MAAT in Lisbon, Makers with a Mission at Somerset House, NEW INC, Future Works NYC, MacDowell, Seoul Art Space Geumcheon, EYEBEAM, and the Foundation of Science and Technology, Portugal.

Joyce Hwang, FAIA, NOMA, is a professor of architecture at the University at Buffalo SUNY, director of Ants of the Prairie, partner at Double Happiness, and a core organizer for Dark Matter U. For nearly two decades, she has been developing projects that incorporate wildlife habitats into constructed environments. She is a recipient of the WOJR/Civitella Ranieri Architecture Prize, the Exhibit Columbus University Research Design Fellowship, the Architectural League Emerging Voices Award, and the MacDowell Fellowship. Her work has been featured by a range of platforms, including MoMA, Brooklyn Botanic Garden, Matadero Madrid, and the Venice Architecture Biennale.

Kyoung Hee Kim, PhD, AIA, is a professor of architecture and director of the Integrated Design Research Lab at UNC Charlotte. She is the founder of EcoClosure, a university spin-off specializing in regenerative technologies by integrating microalgae into building material systems, and a design principal at Hui Kim Design and Architecture (HK DnA). With over eighteen years of industry experience, her work bridges architectural technology, material innovation, and biology. Her research advances nature-based solutions for a carbon-neutral, resilient built environment, focusing on translating scholarly discoveries into practical sustainable technologies that transform the relationship between buildings, users, and ecosystems.

Maria Kuptsova is an artist, architect, researcher, and educator working internationally. In her artistic practice, she develops cyborganic objects, synthetic sculptures, biotechnological interfaces, computational drawings, bio-digital materials, and living architectures. Kuptsova is a PhD candidate and lecturer at the Synthetic Landscape Lab, University of Innsbruck. She received her master's degree from the Institute for Advanced Architecture of Catalonia (IAAC) in Barcelona, and has taught and conducted research at IAAC, the University of Innsbruck, the Bartlett, the DigitalFUTURES conference, and Xi'an Jiaotong-Liverpool University, among others. Her individual and collaborative work has been showcased at the Venice Architecture Biennale, Tallinn Architecture Biennale (TAB), aut. architektur und tirol, VIDEOFORMES Hybrid and Digital ArtS Festival, CYFEST International Media Art Festival, Llum BCN festival, among many others.

Mae-ling Lokko is a designer, architectural scientist, and educator. Her work focuses on the research and development of bio-based construction materials and the integration of next-generation building technologies into low-carbon buildings. Lokko is also the founder of Willow Technologies, a Ghana-based company that upcycles agrowaste into sustainable building materials, and an assistant professor at the Yale School of Architecture where she teaches environmental design and courses on the history and contemporary design of bio-based building technologies. At Yale's Center for Ecosystems in Architecture she directs doctoral research on the whole life cycle development, distributed infrastructures design, and policy around nontoxic, low-carbon materials and next-generation building technologies.

Frank Melendez is co-founder and partner of bioMATTERS, and associate professor at the Spitzer School of Architecture, CCNY, CUNY. He is the author of *Drawing from the Model* (Wiley, 2019) and co-editor of *Data, Matter, Design* (Routledge, 2020). His work has been exhibited at London Design Festival, Usagi NY Gallery, Lisbon Architecture Triennale, and Today Museum. He has curated symposia and exhibitions, including rE-Ordering Architecture in New York and DATA & MATTER in Venice. Melendez has received grants and residencies from Haystack, Somerset House, NYSCA/ Van Alen Institute, the Nature, Art & Habitat Residency in Bergamo, MacDowell, and NEW INC. He holds degrees in architecture from the University of Arizona and Yale University.

Paul Nicholas is associate professor at the Centre for Information Technology and Architecture, Royal Danish Academy—Architecture, Design, and Conservation, and director of the international master's programme in computation in architecture. Nicholas's research develops new methods for architectural design and fabrication based on digital information flows between predictive modeling, registration, and making. He is the founder and lead of the Building with Blue Biomass Global Innovation Network.

Brenda Parker is an associate professor of sustainable bioprocess design at the Department of Biochemical Engineering, UCL. A biochemical engineer by training, her postdoctoral research was conducted within the Algal Biotechnology Consortium at the University of Cambridge. In 2015, she received a Churchill Fellowship to investigate the role of designing for bioremediation. Together with the architect Professor Marcos Cruz, she cofounded the Bio-Integrated Design Lab and postgraduate programme in the subject in 2018 with the aim of developing novel applications of biotechnology for the built environment. She is co-inventor of a patent on cyanobacterial engineered living materials for biomineralization.

Claudia Pasquero is a leading architect and researcher developing climate-active systems that merge biology, computation, and design. She is internationally known for her pioneering work in photosynthetic architecture and synthetic landscapes—built environments that capture CO_2, clean air, and produce biomass. Through her platform PhotoSynthetica™ (a collaboration between ecoLogicStudio, UIBK, and UCL), she has developed algae-based facades, canopies, and pavilions that function as living infrastructures, contributing directly to carbon sequestration and urban air purification. In her book *Biodesign in the Age of AI: Deep Green* (Routledge, 2023), co-authored with Marco Poletto, she explores how architecture can evolve from static form into dynamic, biosynthetic infrastructure.

Marco Poletto is an architect, educator, and innovator based in London. He is co-founder and director of the architectural practice ecoLogicStudio and the design innovation venture PhotoSynthetica, focused on developing architectural solutions to fight climate change. In the past ten years ecoLogicStudio has designed and built several living installations and architectures, demonstrating how microorganisms such as algae can become part of the biocity of the future. Their projects have been exhibited internationally, at Venice Biennale (2025), Louisiana Museum (2024), Hyundai Motorstudio Korea (2023), Centre Pompidou (2022), COP26 (2021), Mori Gallery (2019), ZKM (2019), and EXPO 2017.

Ronald Rael is an architect, author, and professor at the University of California, Berkeley, recognized for his innovative work at the intersection of design, technology, and social justice. His research explores borderwall studies, additive manufacturing, and earth architecture. Rael is the acclaimed author of *Borderwall as Architecture* (University of California Press, 2017), which re-imagines the US-Mexico border through a lens of critique and creative possibility. His visionary work can be found in the permanent collections of the Museum of Modern Art, the Cooper Hewitt Smithsonian Design Museum, the London Design Museum, and many others.

Ehab Sayed, founder and chief Evolution officer at Biohm, is a regenerative design engineer, strategic biomimic, built environment innovator, and public speaker who strives to realize a biomimetic (nature-inspired) future that harmonizes and realigns cultural and natural systems. He founded Biohm to develop nature-inspired biotechnologies aimed at decarbonizing industries through scaling ground-breaking regenerative products and systems. Leading teams of passionate and talented designers, engineers, scientists, and business innovators from around the globe, he is working towards revolutionizing economies to secure the future of humanity without impeding other species' evolutionary pathways—a march toward biosynergy.

Milad Showkatbakhsh, the director of the Emergent Technologies and Design [EmTech] program at the Architectural Association (AA), is a leading expert in bio-inspired advanced generative design, digital fabrication, and robotic construction. His research focuses on the application of evolutionary computation in architecture and urban design, innovative manufacturing techniques, and additive manufacturing to create complex, high-performing architectural assemblies and urban tissues in ecologically sensitive contexts. He pioneers the integration of AI in design automation and develops adaptive building systems and bio-inspired design strategies. Showkatbakhsh's pedagogical approach emphasizes interdisciplinary collaboration and experimental methodologies, directly aligning with [EmTech]'s hands-on, future-oriented curriculum.

Neil Spiller is a contributing editor to *Architectural Design* (AD). A former visiting professor of architecture at Carleton University, Ottawa (2020–22) and visiting professor at Iuav University of Venice (2021), Spiller was previously Hawksmoor Chair of Architecture and Landscape and deputy pro-vice-chancellor of the University of Greenwich, London (2013–18), where he was also dean of the School of Architecture, Design, and Construction, and professor of architecture and digital theory (2010–18). He was vice-dean and graduate director of design at the Bartlett School of Architecture, UCL (2004–10). His architectural design work has been published and exhibited on many occasions worldwide. He is also recognized internationally for his paradigm-shifting contribution to architectural discourse, research/experiment, and teaching.

Mette Ramsgaard Thomsen examines the intersection of architecture and computational design. In 2005, she founded the Centre for Information Technology and Architecture (CITA) at the Royal Danish Academy—Architecture, Design, and Conservation, pioneering research into digital-material relations. Her interests in sustainability and bio-based materials have led to a focus on reparatory practice and systems of return. Her leadership extends to global platforms; she is the Paul Philippe Cret Visiting Professor at the University of Pennsylvania and served as General Reporter for the 2023 UIA World Congress of Architects, advocating for sustainable architecture aligned with the United Nation Sustainable Development Goals.

Michael Weinstock, the founding director of the Emergent Technologies and Design [EmTech] program at the AA, is a seminal figure in digital architecture. His research applies complex systems theory, evolutionary biology, and material computation to explore resilient and adaptive architectural forms inspired by natural systems. Weinstock's pioneering contributions include influential work on biomimetics and natural computation. His current investigations into the ecological implications of building systems and self-organizing structures provide [EmTech]'s deep theoretical foundation. He develops pedagogical frameworks integrating scientific principles with architectural practice, ensuring [EmTech]'s reputation for cutting-edge biodesign and its profound influence on future architectural discourse.

Mark Wigley is professor and dean emeritus at the Columbia Graduate School of Architecture, Planning, and Preservation. His work as an architectural theorist, historian, and critic explores the intersection of architecture, art, philosophy, culture, and technology. Wigley's books include *Buckminster Fuller Inc.: Architecture in the Age of Radio* (2016), *Cutting Matta-Clark: The Anarchitecture Investigation* (2018), *Konrad Wachsmann's Television: Post-Architectural Transmissions* (Sternberg: 2021) and, with Beatriz Colomina, *Are We Human: Notes on an Archaeology of Design* (2016) and *We the Bacteria: Notes Toward Biotic Architecture* (2025), all published by Lars Müller. He has curated exhibitions at MoMA, the Drawing Center, Witte de With Center for Contemporary Art, the Nieuwe Instituut, the Canadian Centre for Architecture, Power Station of Art, the 3rd Istanbul Design Biennial, and the 2025 Milan Triennale.

A D

WHAT IS ARCHITECTURAL DESIGN?

Since its inception in 1930, *Architectural Design* (AD) has stood at the forefront of architectural discourse—a forum where emerging ideas, critical voices, and experimental practices intersect.

The AD of today will explore theory, practice, and culture —*an ecology of architectural ideas*—through critical reflection, dialogue, and action.

AD combines the currency and topicality of a newsstand journal with the rigour and production qualities of a book. We publish four themed issues per year, curated by the AD team and guest-editors committed to probing the architectural culture of cities and regions worldwide.

Visit our website or send us an email:
www.architecturaldesign-ad.com
subscriptions@architecturaldesign-ad.com

We publish in both print and digital formats.

Individual backlist issues of AD are available as books for purchase starting at US$40

PAST ISSUES

VOLUME 95.1

STAGED: ARCHITECTURE FOR PERFORMANCE EXHIBITION AND FICTION

ISBN 978-1-961856-98-1

VOLUME 95.2

ARCHITECTS AND FURNITURE

ISBN 978-1-961856-99-8

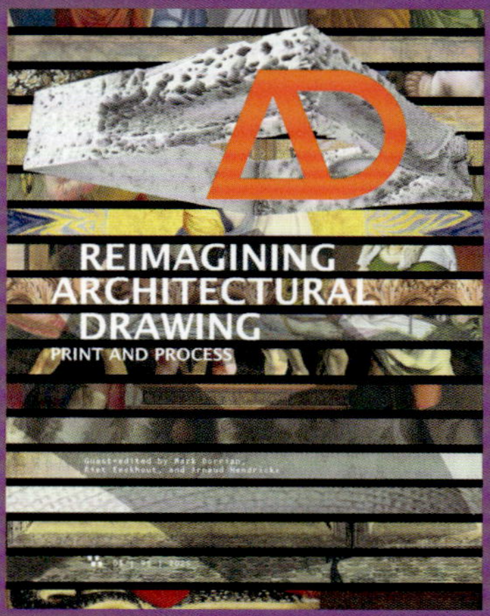

VOLUME 95.3

REIMAGINING ARCHITECTURAL DRAWING: PRINT AND PROCESS

ISBN 978-1-966515-00-5